The Analog Revisited

Questioning the Technical Image in a Digital Age

Edited by
renée c. hoogland
Wayne State University

Series on Critical Media Studies
VERNON PRESS

www.vernonpress.com

In the Americas:

Vernon Press
1000 N West Street, Suite 1200,
Wilmington, Delaware 19801
United States

In the rest of the world:

Vernon Press
C/Sancti Espiritu 17,
Malaga, 29006
Spain

Series on Critical Media Studies

Library of Congress Control Number: 2026937047

ISBN: 979-8-2616-0027-5

Cover art work @ Cynthia Greig

Table of Contents

List of Contributors

Alina Cherry is Associate Professor of French and Associate Director of the Humanities Center at Wayne State University. Prof. Cherry's scholarship focuses on modern and contemporary French and Francophone literatures and cultures, with particular attention to war and memory, space and place, sound studies, and representations of catastrophe in life writing. Her first book, *Claude Simon: Fashioning the Past by Writing the Present* (Fairleigh Dickinson UP, 2016) considered how key stylistic innovations developed by Claude Simon open new ways of understanding the impact of some of the most traumatic conflicts of the twentieth century: the two World Wars and the Spanish Civil War. Her work has appeared in leading journals, including *French Forum, Modern Language Notes, Nouvelles francographies, Symposium, Contemporary French and Francophone Studies: SITES, The French Review*, and *Modern & Contemporary France*, among others. She is currently working on a book project that examines the function of memory in memoirist narratives written in response to four catastrophes that have profoundly marked the twentieth and twenty-first centuries: the Holocaust in *La douleur* by Marguerite Duras, the Rwandan genocide in *L'ombre d'Imana* by Véronique Tadjo, the 2010 Haiti earthquake in *Failles* by Yanick Lahens, and the 2011 triple disaster in Japan in *Fukushima: Récit d'un désastre* by Michaël Ferrier.

Colin Gardner is Distinguished Professor Emeritus of Critical Theory and Integrative Studies at the University of California, Santa Barbara, where he taught in the departments of Art, Film & Media Studies, the History of Art and Architecture and the Comparative Literature program. His most recent monograph is *Chaoid Cinema: Deleuze and Guattari and the Topological Vector of Silence* (Edinburgh University Press, 2021), which explores the use of sonic drop-outs in sound films in order to explore different organizations of chaos (Chaoids) that underlie the surface plane of narrative. This builds upon his previous book, *Beckett, Deleuze and the Televisual Event: Peephole Art* (Palgrave Macmillan, 2012), a critical study of Samuel Beckett's experimental work for film and television and two books for Manchester University Press's 'British Film Makers' series: *Joseph Losey* (2004) and *Karel Reisz* (2006). He has also co-edited two anthologies with Patricia MacCormack (Anglia Ruskin University): *Deleuze and the Animal* (2017, for Edinburgh University Press) and *Ecosophical Aesthetics: Art Ethics and Ecology with Guattari* (2018, Bloomsbury Academic).

Ted Hiebert is an interdisciplinary artist and theorist. His work examines the relationships between art, performance, and technology with a particular focus on the absurd, the paradoxical and the imaginary. He is the author and editor of several books including, most recently, *Photographing Ambiguity* (University of Toronto Press, 2025), *Postrational Visuality* and *More Postrational Visuality* (co-edited with Duncan MacKenzie, Noxious Sector Press 2025 and 2026). Hiebert is Professor and Chair of the School of Image Arts at Toronto Metropolitan University. www.tedhiebert.net.

renée c. hoogland is a professor of English Literature and Culture at Wayne State University in Detroit. She is the author of 40+ essays and book chapters and three monographs: *A Violent Embrace: Art and Aesthetics after Representation.* University Press of New England, 2014; *Lesbian Configurations.* Polity Press & Columbia University Press, 1997; *Elizabeth Bowen: A Reputation in Writing.* New York University Press, 1994. hoogland served as the editor in chief of the ten- volume McMillan Handbook Series on Gender and is currently the editor of *Criticism: A Quarterly for Literature and the Arts.* She is working on a book about photography, entitled, *The Other Side of Nowhere: Thoughts on Contemporary Photography.

Aaron Jaffe is Frances Cushing Ervin Professor at Florida State University, USA. He is the author or editor or co-editor of seven previous books, including *The Way Things Go: An Essay on the Matter of Second Modernism* (2014), *Spoiler Alert: A Critical Guide* (2019), *Understanding Flusser, Understanding Modernism* (2021) and Flusser's *Thinking Further* (2026). He's working on a book on modernism and media theory.

Paula J. Massood is Dean of the School of Visual, Media, and Performing Arts at Brooklyn College, CUNY. She is also Professor of Screen Studies at Feirstein Graduate School of Cinema at Brooklyn College, and on the doctoral faculty in the Film Studies Certificate Program and Program in Theatre at The Graduate Center, CUNY. She is the author of *Black City Cinema: African American Urban Experiences in Film* (Temple UP, 2003) and *Making a Promised Land: Harlem in 20th-Century Photography and Film* (Rutgers UP, 2013), editor of *The Spike Lee Reader* (Temple UP, 2007), and coeditor of *Media Crossroads: Intersections of Space and Identity in Screen Cultures* Duke UP, 2021), *The Routledge Companion to American Film History* (Routledge, 2025), and the "Precarious Mobilities," special issue of the journal, *Feminist Media Histories* (U of CA Press, 2021).

Elena Past is Interim Dean of the Irvin D. Reid Honors College and Professor of Italian at Wayne State University in Detroit, Michigan. In 2022, she was a Rome Prize Fellow at the American Academy in Rome (2021-22) and a Fulbright U.S. Scholar (2022). Her books include *Italian Ecocinema Beyond the Human* (2019), winner of the Modern Language Association's Howard R. Marraro Prize; *Methods of Murder: Beccarian Introspection and Lombrosian Vivisection in*

Italian Crime Fiction (2012); and the co-edited collections *Landscapes, Natures, Ecologies: Italy and the Environmental Humanities* (with Serenella Iovino and Enrico Cesaretti, 2018) and *Thinking Italian Animals: Human and Posthuman in Modern Italian Literature and Film* (with Deborah Amberson, 2014). With Marco Armiero and Roberta Biasillo, she co-edits EditPress's new book series in Environmental Humanities. Her research interests and publications include work on environmental media studies, critical animal and critical plant studies, posthumanism, and Italian film studies.

Judith Roof retired from Rice University as the William Shakespeare Chair in English. She is the author of eight monographs and essays on topics including narrative theory, sexuality, Hollywood cinema, DNA, the shift from analogue to digital, psychoanalysis, gender, film theory, hoaxes, *The Big Lebowski*, nerds, viagra, James Bond, feminist criticism, protozoa, systems theory, critical legal studies (she is an attorney), and the work of such authors as Samuel Beckett, Harold Pinter, Marguerite Duras, Virginia Woolf, Percival Everett, Richard Powers, Nicole Brossard, David Hare, Simon Gray, Tom Stoppard, and Rabelais. She has also edited or co-edited collections of essays on dramatic criticism, the "oddball archive," feminist criticism, and psychoanalysis. Monographs include *A Lure of Knowledge: Lesbian Sexuality and Theory, Come As You Are: Narrative and Sexuality, Reproductions of Reproduction: Imaging Symbolic Change, All About Thelma and Eve: Sidekicks and Third Wheels, The Poetics of DNA, What Gender Is, What Gender Does, The Comic Event,* and *Tone: Writing the Sound of Feeling.*

List of Figures

Introduction

In *Towards a Philosophy of Photography* (1983), Vilém Flusser defines images as "significant surfaces" that "make comprehensible" something "out there" in the four dimensions of space and time; images paradoxically make the world comprehensible by reducing it to a two-dimensional abstraction. As such, images operate in the realm of magic, mediating between humans and a world in which we merely "ex-ist."[1] Yet, the rise of what Flusser terms the "technical image," in its endless reproducibility and ubiquity, threatens to transform images from "maps" into "screens" that we no longer "decode" but simply accept, so that our lives become a "function of their own different own images" and "imagination turns into hallucination."[2] Flusser died in 1991, before the digital revolution would irreversibly overwhelm the domain of images, but, like Jean Baudrillard in "The Precession of Simulacra" (1981), he anticipates an increasingly visualizing culture in which digital technologies fundamentally change the image and its function, as humans no longer distinguish between "reality" and representations of reality.

In contrast, Jacques Rancière, in *The Future of the Image* (2003), rejects the idea that contemporary reality is "devoured" by media and synthetic images, such that there is nothing but images. Refusing to conflate a "certain idea of fate" with a "certain idea of the image"—a conflation that defines our current cultural climate in decidedly "apocalyptic" terms—he instead asks: "[A]re we in fact referring to a simple, univocal reality? Does not the term 'image' contain several functions whose problematic alignment precisely constitutes the labour of art?"[3] Rather than regarding technical images as "screens" that turn imagination into hallucination, Rancière posits that images, as "aesthetic acts," are "configurations of experience that create new modes of sense perception."[4]

This collection of essays aims to explore the complexities of the technical image in the wake of the digital turn. While the digital image has, in effect, effaced the necessity for a pre-existing "reality," the presumed indexical function of the photographic image has by no means disappeared—technical images continue to organize, if not generate, our shifting modes of perception. How do we reconsider analog photographs, given that they can and are likely

---

[1] Vilém Flusser, *Towards a Philosophy of Photography* (London: Reaction Books, 1983) 8.

[2] Ibid. 10.

[3] Jacques Rancière, *The Future of the Image* (London: Verso, 2009) 1.

[4] Jacques Rancière, *The Politics of Aesthetics* (London: Continuum, 2006) 9.

to show up on a variety of digital platforms? How do we re-view technical images that make the invisible visible, given that current imaging technologies generate new visualities as such? How does the digital force us to reconceptualize analog photography?

Individual chapters engage with a range of technical images—past and present—approached from widely different critical and/or theoretical perspectives. Topics include visibility versus visuality; object-orientation; human-nonhuman relations; and the affective, haptic, and other sensorial aspects of technical images in their aesthetic and political dimensions.

When all the photographs have been taken, as Jean Baudrillard prophesized, what will remain is the experience of taking a picture, the guarantee of an incommensurable moment that will forever defy representation because it is itself the condition for representation in the first place. The ubiquitously realized technical image is thus both the marker of immersive hyperreality and an exit through which to explore the embodied condition of post-representational experience. In "Exercises for seeing the image differently," Ted Hiebert takes up the challenges of engagement and performance as possible ways to rethink what cameras can do, staging playful experiments to create moments specifically for the image. His participants wear chairs on their heads, cross their eyes for a picture, and shine laser pointers at the camera—in each case challenging the directionality of the digital gaze and insisting on the (analog) presence of subjects, bodies, and experiential moments as foundational to the operation of the photograph.

In "From Analogue Facialities to Digital Probeheads: Deterritorializing the Photographic Composites of Francis Galton and Nancy Burson," Colin Gardner applies Gilles Deleuze and Félix Guattari's twin axes of faciality – the white wall of *signifiance* and the black hole of subjectification – to the generic and racist analogue composite portraiture of Francis Galton at the turn of the twentieth century, compared to Nancy Burson's digital equivalent, which breaks down the *signifiance*/subjectification structure of the face through a deterritorializing and transforming probehead. Although the latter never fully escapes the dominance of the face in terms of structure and signification, it turns them against themselves as an affective embracing of different temporalities and races (Burson is most famous for her *Time* magazine cover that morphs Trump and Putin). Burson achieves her greatest success in her 1990s androgynous *He/She* series, which are all the more powerful because we expect them to be digital composites (and therefore "artificial") when, in fact, that are purely indexical.

Elena Past, in "Bodies of Work: Ferrania Film's Analog Legacies in the Anthropocene," takes up the Ferrania film manufacturing facility in Liguria, Italy, which was Italy's Kodak. For much of the twentieth century and into the twenty-first, they made photosensitive analog film stock for movies, photography, and X-rays. In the 1960s, the factory was acquired by Kodak competitor and American giant 3M, and in the early 2000s, like many analog production companies, they turned out the lights. Past's research into the Ferrania factory's history is rooted in the Environmental Humanities and in the material impacts of the production, distribution, and preservation of analog film. This chapter examines the role of the factory's female laborers, the "filmiste," by crossing factory histories with photographic evidence that circulates today on a lively Facebook group where former workers share memories of factory life. Photographs on Facebook allow Past to think bodies across decades and across media. Making photosensitive film stock was a trans-corporeal affair, in which female bodies and more-than-human bodies shared sensitivities, for example, to chemicals, dust, and radiation. Photochemical aging, visible in old photographs posted on Facebook, recalls the aging bodies of a slowly disappearing labor force, and points to some of the industrial exposures that may have shortened laborers' lives. Though the digital in theory stabilizes the photographic aging process, the Facebook page also serves as a site where memories are shared and communities are built. Repetitive factory labor in many cases constituted a space of affective intimacy, an experience that endured long after jobs concluded. The interwoven threads of these stories, written across analog and digital media, create a patchwork view of analog's legacies in the twenty-first century: multiple generations of bodies, materials, and labor woven into the fabric of memory and matter.

Judith Roof's "On Posing" explores the different ways non-professional or "everyday" people who pose for photographs and models who pose professionally may both know (or not) and adjust to the differences they may perceive between analog and digital photographic technologies. Such differences in the technologies as shutter speed and length of pose, lighting, time of development, the alterability of the photographs and the photographs' likely survival contribute both consciously and unconsciously to posing practices depending on the contexts, knowledge, and purposes of the photographs.

In "Compelling Abstraction: Cynthia Greig's Digital Transformations of Art Space," renée c. hoogland addresses the loss of the aura of the artwork, which has been bemoaned since Walter Benjamin announced its disappearance in the age of mechanical reproduction. Decades later, Roland Barthes restored the aura to the analog technical image by endowing it with the power to produce a trace, the co-presence of a subject and a surface reality. The digital image is often assumed to have destroyed this last remnant of an "authentic" reality by throwing representation per se into crisis. hoogland argues against this narrative of loss. Instead, she explores ways in which the digital image actualizes visibility where none was before, in particular in the realm of aesthetic production and its reception, in the domain of art in its "worlding" functions. The chapter's focus is on the work of Michigan artist Cynthia Greig, especially two sets of digital photographs shot in art galleries and museums across the US and Europe that fundamentally reorganize the relationship between the subject and object of artistic visuality and the field of aesthetic perception as such. "Threshold" comprises a series of large-scale photos of people looking at the walls of museums from which the artworks have been digitally erased. "Gallery Horizons" consists of close-ups of the space where the floor of the gallery meets the wall, typically invisible spaces that are digitally transformed, abstracted to horizon lines, thus actualizing deep landscapes that emerge only in the perceptual encounter, in the event of looking—elsewhere.

Aaron Jaffe's chapter "Connecting Modernism and New Media Theory: Bruno… Pound. Flusser... Joyce" tracks the intersection of modernist studies and new media theory, focusing on Flusser's concept of the technical image. It traces the intellectual lineage of this concept through figures like Giordano Bruno, Ezra Pound, and James Joyce, highlighting their shared interest in a decentered, infinite universe and the role of the metamedium in shaping perception and critical practice. The analysis examines how these modernists, viewed as technologists of the risk-medium, challenged traditional notions of form and meaning, embracing risk and incoherence. The chapter further connects this modernist approach to Flusser's later work, emphasizing the significance of medium orientation in a world saturated with technical images. The chapter uses Samuel Beckett's chapter title, "Dante... Bruno. Vico.. Joyce," as a framework to structure this intellectual genealogy.

In "Scratching the Surface: Irma Vep's Analog and Digital Histories," Paula Massood traces the character's earliest appearance in Louis Feuillade's silent serial, *Les Vampires*, to her later iterations in *Irma Vep* (Olivier Assayas, 1996) and again in the HBO limited series, *Irma Vep* (Olivier Assayas, 2022). The character of Irma Vep has functioned not only as a symbol of criminal and sexual transgression—a shape-shifting fugitive from the law—but also as a springboard for examinations of filmic and digital materiality. Beginning in an analog and highly combustible nitrate form, Irma metamorphized over more than a century's worth of technological shifts and industrial globalization to become a digital and transnational container of the past and present, of history, of representation, and of the very ontology of cinema. In the 1996 film, for example, the role most famously played by Musidora is performed by Hong Kong actress Maggie Cheung, Assayas's casting choice thus openly questioning definitions of national cinema while also—in an ending sequence shot in 35mm—referencing film's material base. Likewise, the most recent streaming serialization further expands questions of nationalism and materiality by looking both backwards towards the historical roots of the character and

French cinema and forwards, situating the narrative in a digital and global streaming present dominated by the Marvel Cinematic Universe and starring a Swedish actress most well-known for her work in American film. This chapter argues that such shifts, makeovers, reboots, and remakes provide fertile ground for understanding cinema's shift from the nitrate film strip to the digital stream. The focus is on three related moments featuring Irma Vep—from 1915, 1996, and 2022, respectively—to illustrate early film's persistent presence in a digital world and to argue that the analog is no more "real" than its digital counterpart. Do filmic images, as Vilém Flusser argues, "absorb the whole of history and form a collective memory going endlessly round in circles"?[5] Alternatively, do they suggest more of Roland Barthes argument regarding cinema's "protensity"; its continuity and flow, the latter an apt term as well for serialized television.[6] Finally, do they prove the "death of cinema," as many film historians have feared, or do they suggest an expansion of the visual predicted by André Bazin in his writing on new technologies and television?

In "Making Catastrophe Visible: Photography, Time, and Matter," Alina Cherry explores the multifaceted role of analog photography in the aftermath of Japan's 3/11/2011 disaster, a compound catastrophe consisting of an earthquake, tsunami, and nuclear accident. Specifically, she examines how photography engages with the emotional, temporal, and ethical complexities of catastrophe through three interrelated lenses: the recovery and restoration of tsunami-damaged photographs; the challenges of visually representing disaster; and the visualization of radioactive contamination. Drawing on theoretical insights from theorists such as Roland Barthes, Susan Sontag, and John Berger, as well as the practices of artists like Thierry Girard, Minato Chihiro, Shimpei Takeda, and Yoi Kawakubo, the chapter foregrounds how photography resists oblivion, makes the invisible visible, and fosters collective memory.

Focusing on analog photography, Cherry highlights how the material fragility and persistence of this medium contribute to its ontological weight. Volunteer-led efforts to salvage water-damaged photographs serve not only as acts of remembrance and healing but also as communal affirmations of existence amid loss. At the same time, cameraless practices that use contaminated soil to expose film make visible the unseen and enduring threat of radiation, challenging anthropocentric notions of vision and agency. These autoradiographic images, shaped by nonhuman forces, function as both ecological testimony and aesthetic resistance. Ultimately, the chapter argues that photography's deeper significance lies not merely in documenting destruction, but in revealing what disasters

---

[5] Vilém Flusser, *Towards a Philosophy of Photography.* (Chicago: University of Chicago Press, 1983), pp. 19–20.

[6] Roland Barthes, *Camera Lucida.* Trans. Richard Howard. (London: Flamingo, 1984), p. 90.

leave behind, erase, or render invisible. By engaging with catastrophe across material, emotional, and philosophical dimensions, photography emerges as a vital medium for confronting the limits of representation and bearing witness to loss, trauma, and environmental devastation.

Individually and collectively, the chapters in this book take up the challenge of the technical image in its various permutations—analog, digital, moving and still—and reach a variety of conclusions about its operations in the digital age. As such, the chapters shed light on the various ways in which the technical image operates, both with regard to specific sets of images per se, and in the ways in which it functions within wider theoretical contexts.

Chapter 1
Exercises for Seeing the Image Differently

Ted Hiebert
Toronto Metropolitan University

Abstract

What if we stopped using cameras to make sense of the world and started using them to make it stranger? This chapter challenges the quantifiable, data-driven ideology of 21st-century digital image culture by proposing a series of absurdist exercises—performative, participatory, and pedagogical—that treat photography as a medium of ambiguity and creative, analog, resistance. By wearing chairs on our heads and shining lasers at surveillance cameras, it is possible to engage new ways of seeing and imagining the world around us, and one that disrupts normalized technological programming. The result is a suggestion of ways to reclaim photography, not as a practice of documentation, but as an ecological and collective act of world-building.

Keywords: Photography, Ambiguity, Performance, Creativity, Digital Culture, Art.

***

This essay looks at photography as a metaphor for knowledge practices in a technological culture, arguing that a relational examination of the medium of photography can shed new light on some of the dominant ideological tendencies of the 21st century. In making this suggestion, I align myself with thinkers such as Vilem Flusser, François Laruelle, Amanda Boetzkes, micha cardénas, Katherine Behar and others, arguing for uses of images that can highlight and catalyze more ambiguous ways of relating to the world. In my ways of thinking the image, I foreground ambiguity as a way to cultivate more conscientious, ecological, and creative relationships within the technological ecosystem of contemporary life, specifically because the stakes of ambiguity are experiential and performative rather than documentary or instructional. This essay thus critiques the idea that images are best deployed to verify or authorize a world outside of us, and proposes that such (quantifiable) ways of thinking—while rooted in the historical drive towards technology and data (what Arthur Kroker calls "the will to technology"[1])—have become so pervasive as to have become a dominant form of ideological bias in the 21st century. Against the ubiquity of data consciousness, this essay proposes a series of exercises designed to cultivate an embodied encounter with digital living, not against the clouds of images, but within them—understanding photographs not as benchmarks of the real but as ambiguous constructions of the present and future imaginary. I propose that images—despite appearances to the contrary—were never really meant to make sense of the world but rather to help make it stranger.

What follows is a series of examples that I have been developing over the last 10 years, designed both as pedagogical tools and as participatory art projects that I produce with students and artistic communities.[2] What they share is an attempt to approach the idea of experiential ambiguity, using the camera to focus on what the camera can never quite capture: the nuances of lived experience. That such a project deploys the tools of creative proposition is not an argument against the authenticity of the moment but rather a proposition anchored in a desire to see and understand photography differently—not as a practice of documentation but as one of calling representation into question. I see it as an analog counterpart to digital thinking, following Flusser and others who claim that the most important effect of the image is not the way it reshapes vision but the way it reshapes ideology and thought itself. In this spirit, these experiments are less about making new kinds of pictures and more about proposing things one might try oneself—or in my case with students—if for no other reason than to participate in the act of using photography to make the world a little bit stranger, a little more down to earth, and perhaps in the process making what cameras do a little more ambiguous.

I would offer that what these exercises have in common is an interest in a form of photography that seeks to disavow the technical project of representation and instead focuses on the performance of what can never be photographed—the ambiguities of lived encounter with individual worlds. To use the camera to multiply the world— and in doing so prove that the world can be multiplied, that reality is not self-evident, and that entanglements of bodies and images and imaginations are the ambiguous norm. To refuse the idea that quantification alone leads to valuable knowledge and to insist on preserving some of the more

---

[1] See Arthur Kroker, *The Will to Technology and the Culture of Nihilism: Heidegger, Marx, Nietzsche* (Toronto: University of Toronto Press, 2015).

[2] A complete archive of these projects, along with participant images, can be found online at (http://www.tedhiebert.net). These are living projects and additional contributions are always welcome.

ambiguous spaces of lived reality—incommensurabilities and absurdities and failures and irrelevances and conspiracies and contaminations and falsities and dreams. Not benchmarks of the real but ambiguous constructions of present and future imaginaries. It's not about what the camera does anymore. It's about what we do with it—or to it.

exercise 1: wear a chair on your head

In the novel *Au Plafond* (On the Ceiling) by the French writer Éric Chevillard, the main character wears a chair upside-down on his head.[3] It was, at first, to correct his posture—imagine how straight one would have to walk to keep it balanced! But the man grew to like it, and as such it became about more than correcting his posture—it helped him to correct his imagination of himself too. It's also a rechoreographing of the relationship to a chair. The problem with chairs is that they expect us to sit on them, they make us face forward, they keep us immobile and aligned with the other bodies in the room. As such, they dictate how we use them. But if we consider chairs as metaphors for all technologies, then it matters a great deal to find different ways to rescript this relationship. To relate to a chair, one must prove that one is an active and attentive partner in the engagement, even if it seems silly to do so.

To seriously consider the idea of wearing a chair on one's head, one needs to move past the idea that chairs are designed only to be sat upon. In so doing, one also might realize that the regular design of the chair not only dictates how the chair itself is used—it dictates how we ourselves use it too. In Chevillard's story, the chair serves as a prosthetic to correct the character's poor posture. In the world outside of the book, such an action might be thought of as a way to engage with the world a little more creatively.

Guidelines: Take a photograph of somebody wearing a chair upside-down on their head. Whether the chair is a hat, a performance prop, or something else, realize that in the process of staging this picture, you have re-scripted the possible uses of the chair and the possible methods for human interaction. Consider staging your intervention in a public location.

3 Éric Chevillard, *On the Ceiling*, Jordan Stump, trans. (Lincoln: University of Nebraska Press, 2000).

Fig. 1.1: Maura Bennett, *On the Ceiling*, color photograph, 2018.

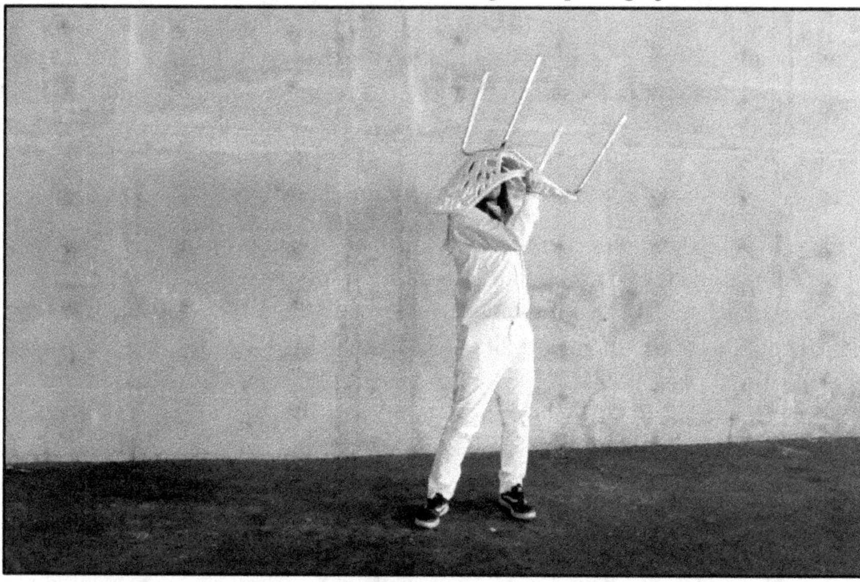

From the project *On the Ceiling*. Participant image: Maura Bennett, 2018.

I think of Francois Laruelle and his non-philosophy project, where the point of philosophy is not to simply think differently but to "exhaust itself as an immanent practice rather than as a program."[4] Necessary, then, is to understand and engage with the programs as they are currently running—the question of whether these programs are algorithmic or material, I would contend, is perhaps not the point. Chairs have programs too, and they run us most of the time. To engage a chair non-philosophically (in Laruelle's sense) would be to reverse this relationship in some way—to ask, as Laruelle does, how it can be put to use differently.[5] And so too for the camera; for Laruelle, performance is key, exceptions are to be coveted, and the endgame of philosophy (and photography) is immersion in a sustainable fiction of its own making. Laruelle calls it "photo-fiction".[6]

I believe that this is where the work of the artist is important, since, as new media artist and activist micha cárdenas eloquently puts it, the work of the artist is to "create believability," especially when dealing with non-artistic or

---

[4] François Laruelle, *The Non-Philosophy Project: Essays by François Laruelle* (Candor: Telos Press Publishing, 2012), 219.

[5] Laruelle, *The Non-Philosophy Project*, 207.

[6] Laruelle *Photo-Fiction: A Non-Standard Aesthetics* (Minneapolis: Univocal Publishing, 2012).

real-world contexts.[7] Cárdenas calls a form of "transreal aesthetics,"[8] thinking in the context of a contemporary context that is becoming increasingly and explicitly technological and in which the boundaries between the real and the virtual—indeed between the analog and the digital too—have been fragmented beyond clear recognition. Importantly, instead of trying to reconcile this fragmentation, Cardenas suggests looking to exploit it as a way to "use reality [itself] as medium." [9]

Laruelle calls it an identity that is "performed without an act of performance," "a program-without-programming or at the very least a program for de-programming." Cardenas simply calls it "reality-flexing," a concept that I think becomes proportionally more radical as it becomes integrated beyond the act of performance as an ambition for a new form of life, simply called "living."[10] Cardenas is an activist dealing with much heavier issues than chairs, but I think a lot of what she offers scales to the more modest and mundane moments of living. Even something like wearing a chair on one's head. Certainly, it is an absurdist act performed for the camera, but it is also performed by real people having a real experience while doing so—they have been charged with stepping outside of the scripts of chairs for the sake of using photography differently. I would call it an act of non-choreography, a challenge to the behavioral scripts that mark technological engagement by insisting on a creative proximity to the performance of virtual interaction. Against the scripting of technological behaviors, an absurdist choreography of engagement.

exercise 2: cross your eyes for the camera

In his novel *A Journey to Ixtlan*, Carlos Castaneda describes a conversation with an indigenous sorcerer who teaches him to see the world differently. The technique "consisted of gradually forcing your eyes to see separately the same image. The lack of image conversion entailed a double perception of the world ... which the eyes were ordinarily incapable of perceiving."[11] Seeing the world with crossed eyes, for Castaneda, was a first step towards a larger understanding of vision, multiplied out of the linearity of bifocal sight.

In her essay "Feeling with your eyes," the artist and activist Serena Kataoka takes this notion one step further, arguing that Castaneda's notion of cross-

---

[7] micha cárdenas, *The Transreal: Political Aesthetics of Crossing Realities* (New York: Atropos Press, 2011), 28.

[8] Ibid, 23.

[9] Ibid, 35.

[10] Ibid.

[11] Carlos Castaneda, *A Journey to Ixtlan: The Lessons of Don Juan* (New York: Simon and Schuster, 1972), 72.

eyed vision can also be applied to social and political culture, allowing one to hold multiple—and sometimes conflicting—perspectives in one's mind at the same time.[12] In Kataoka's view, one might see in two different ways out of two different eyes—not simply a difference in optics but potentially one with ideological implications, quite literally allowing for new ways of thinking to emerge. Or proposing the possibility of seeing multiple versions of the world, held together by blur instead of by the usual tropes of focused singular vision.

Guidelines: With Castaneda and Kataoka in mind, *Cross-eyed Visions* is a visual response project that asks participants to meditate on the idea of seeing the world differently. To contribute to this project, take a portrait-style photograph of someone with their eyes crossed. The image stands as a metaphor for seeing in more than one way at a time—inviting those who view it to consider that there are multiple ways of perceiving the world.

Fig. 1.2: Ben Rockhold, *Cross-eyed Visions*, color photograph, 2018.

From the project *Cross-eyed Visions*. Participant image Ben Rockhold, 2018.

[12] Serena Kataoka, "Feeling with Your Eyes: Petroculture and *Ice Follies*," in Ted Hiebert, ed., *Plastic Blue Marble—Catalyst: Amanda Boetzkes* (Seattle: Noxious Sector Press, 2016), 101-102.

Sometimes moments invite being seen with crossed eyes—marked by a kind of dissatisfaction with regular appearances, and a need to make vision excessive—twice the world means no singular identities, no fixed or static objects, thus the possibility for different ways of seeing. It makes no sense to just try to re-focus; focus was the problem. Sometimes two versions are better than one.

I think of Paul Virilio's notion of the "vision machine" that makes human sight accountable to technological standards of assessment rather than the other way around. In this case it's the emphasis on precision focus and high resolution—optically and ideologically. For Virilio, this is also a clear symptom of the dominance of the digital over the analog—and the photograph as exemplary metaphor for the technological domination of the human spirit, stealing not just minds but souls.[13]

This because, for Virilio, any time a technology begins to exceed human capacity (which he theorizes as the dominant trend of the 20th century) it replaces humanly-observable standards with new computational criteria thought more efficiently, more precisely, more objectively and thus less subject to human spirit or nuance.[14] The standards according to which humanity thus becomes subject are literally not observable to humans themselves without the enhanced capacity of technological systems. "The watching gaze has long since ceased to be that of the artist or even the scientist, but belongs to the instruments of technological vision."[15] I share his feeling that photography provides an anticipatory logic for the rise of computational thinking.

I might call it a form of what the philosopher and art historian Amanda Boetzkes calls "ecologicity," the act of "attuning vision to an ecological reality."[16] For Boetzkes, art—at its best—holds the capacity to engage in its present contexts without violating them, a form of Irigaray's "recessive ethics," here deployed towards displacing the colonial obsession with the domination of nature (or, one might offer, Virilio's obsession with the domination of the digital) in favor of an engagement with what it means to be situated in relation to that which one does not or cannot know. In this, Boetzkes seeks potential ways to develop new forms of environmental conscience, built on the idea of thinking art and technology from an explicitly ecological posture.[17] Now, to be fair, Boetzkes is thinking about ecologicity in environmental terms, and her projects are more

---

[13] Paul Virilio, *Politics of the Very Worst* (New York: Semiotext(e), 1999), 54.

[14] Ibid., 86.

[15] Paul Virilio, *The Information Bomb,* Christ Turner, trans. (London: Verso, 2000), 57.

[16] Amanda Boetzkes. "Ecologicity, Vision and the Neurological System," in Heather Davis and Etienne Turpin, eds., *Art in the Anthropocene: Encounters Among Politics, Aesthetics, Environments and Epistemologies* (Ann Arbor: Open Humanities Press, 2015), 276.

[17] Boetzkes. "Ecologicity, Vision and the Neurological System," 276.

ecologically oriented than my own. But I think the idea of being attentive to context is important, and the idea of thinking photography ecologically is worth exploring. To think of images not as objects but as environments resituates the stakes of the question, and makes clear the dominant expectation of clear articulated vision.

And so perhaps there is a workaround to the challenge of precision vision and digitally informed context—a cross-eyed workaround in which the photograph is still clear but the model sees us twice—once out of each crossed-eye. To consider this possibility is to entertain the idea of the image apparatus as a relational device, giving the camera the power to not resolve nuance but rather to hold space for what it itself cannot actually see—the world that appears in two ways at once. A focused image that cultivates sensitivity to blur. I do not see the person in the picture in the same way that they see me. Perhaps with images such as these, it's best not to look with precision at the picture but to cross one's own eyes in an act of solidarity with the project of looking and thinking differently.

exercise 3: shine a laser pointer at the camera

In his book *Towards a Philosophy of Photography*, the philosopher Vilém Flusser suggests that technology has a tendency to overwrite human ways of understanding the world. He likens the process to magic—replacing human experience and knowledge with a technical likeness we believe even more than we believe our own thoughts, feelings, or intuitions.[18] We see this in medicine, where bodies are increasingly understood through testing and genetic codes; we see this in social living, where identities are increasingly represented by data profiles; we see this in knowledge, where databases replace libraries, and information circulates at the speed of access. Under the regime of this sort of technical imagination, we risk becoming simply another source of information for algorithmic systems of computational learning and knowledge.

For Flusser, the only appropriate response to this technological context is to attempt to speak back to the systems that code our ways of knowing and behaving, a game of tag in which we are only human to the extent that we can understand how to remix, respond or re-imagine the ways in which technology creates and represents the worlds we inhabit.[19] It is less a revolution than an attempt to insist on a relationship with the technologies that monitor us as we move through life.

---

[18] Vilém Flusser, *Towards a Philosophy of Photography* (London: Reaktion Books, 2000), 17.

[19] At stake for Flusser is the question of human freedom. Ibid, 80-82.

Guidelines: To participate in this project, take a photograph of somebody shining a laser pointer at the camera. When aimed just right, this will cause a lens flare that can be treated as an artistic element. Consider staging your intervention in a public location. Consider using a publicly available surveillance camera, web camera, or traffic camera as your target. You should be able to see what they see on any mobile device, thus confirming that your intervention has been seen.

Fig. 1.3: Heather McCallister, *Laser point tag*, color photograph, 2019.

From the project *Laser Pointer Tag*. Participant image Heather McCallister, 2019]

The future of photography will be found in the dark—but not the kind of darkness that results from simply turning out the lights. Photographic darkness is more like the silence produced by white noise: a darkness produced by overexposure, an overshadowing whose result is to port material vision into a hallucination produced by technical systems. This kind of darkness is the kind that is impossible not to see—it is a ubiquitous darkness that circulates under the sign of technical impartiality. The camera is the original black box, after all, and thus a forebear of the complex technical and intelligent systems of today.

As Vilem Flusser so aptly puts it: "All apparatuses (not just computers) are calculating machines and in this sense 'artificial intelligences', the camera included."[20] And for me, Flusser can be such a perfect foil for this line of thought because I take him to believe, first and foremost that the primary objective of cameras—and thus of all technical systems—is not simply the production of images but the actual production of ideology: "Their intention is not to change the world but to change the meaning of the world."[21]

But if that's true, then cameras change the way we think—literally. It's important not to gloss this point—if cameras change the way we think then they are not representational technologies but productive ones, engineering new technical modes and standards of apprehension. Thought, for Flusser, is inadequate as a means of resistance since thinking is itself the actual material medium used by the camera.

The trick then, is not to think about images differently, but to think differently about technical thought itself—one version of which might be to implement what artist and theorist Katherine Behar calls "decelerationist aesthetics" as a way to survive the ubiquitous systems of technical capture by finding ways to create pockets of slowness in an ever-accelerating world.[22] It could also be seen as a form of counter-magic—thinking about technical ideology not as an information system but as a context for engagement. It could also be seen as a new form of digital darkness—one that operates through over-exposure. I think of Behar's concept of "Botox ethics," where a Botox injection—literally or metaphorically—has the capacity to provide an emotional mask that hides a certain form of interiority, decelerating the system of public visual judgment by offering instead an image numbed in advance.[23] For me, there is something sympathetic here to the project of pointing lasers at cameras. A certain kind of opacity—maybe even a kind of digital darkness.

I used to lament the loss of the darkroom as the material counterpoint to the optical emphasis of the photographic medium. For me, the darkroom was a site of blindness, characterized by exactly the opposite of sight—as a photographer develops film in the dark, waiting for the images to emerge, navigating by touch and memory rather than by optical markers of space. The idea that this experience could be produced is intriguing—like a white noise machine, a laser pointer for the camera.

---

[20] Ibid, 31.

[21] Ibid, 25.

[22] Katherine Behar, *Bigger than You: Big Data and Obesity* (Goleta: punctum books, 2016), 2-3.

[23] Katherine Behar, ed., *Object-Oriented Feminism* (Minneapolis: University of Minnesota Press, 2016), 139.

And so maybe digital darkness isn't the context created by pervasive machine magic but just the opposite. An intentional decision to live in the dark, which means to feel one's way around even if one can still see. To live *as if* in the dark, such as to remain reminded of the magic wash of intelligibility. If intelligibility—as a marker of technical thought or algorithmic virtuosity—is a magic spell, to break the spell one needs to relinquish a reliance on making sense. Against the high-resolution representation of truth, a purposeful alignment with the ambiguity of a world unseen.

conclusion: wear a tinfoil hat

By way of a conclusion, I want to share one last project, though I'm not really going to talk about it in any depth. It's a simple one—photographing people wearing tinfoil hats. It's my conclusion because I think it's a perfect metaphor for the idea of thinking about technical images in a digital age—the idea that critical thought can create a sort of proverbial tinfoil hat that shelters us from some of the pervasiveness of digital living. Or the idea that resistance is probably futile, but sometimes can still be fun and maybe even create pockets of community along the way. Either way, a tinfoil hat seems like an interesting metaphor—a shield, a performance prop, a fashion accessory, a moment of critical resistance, real or imagined—a signal to the signals that circulate around us and in some ways bind us together.

Fig. 1.4: Brandon Kan, *Tinfoil hats*, color photograph, 2021.

From the project *Tinfoil Hats*. Participant image Brandon Kan, 2021]

It's ambiguous, of course, which seems about right as a qualification of what it means to live and think at a time when ambiguity is an offense to the technical project of full-spectrum information and high-resolution living. When all the images have been taken—when, as Jean Baudrillard prophesied there is nothing left to document—what remains is the collective solidarity of lived experience which stands in ambiguous defiance of the project of full spectrum capture for the simple reason that realities are created not given, and ultimately not accountable to the evidence provided by images but rather by the possibilities they engender. As a provocation in the context of this panel, I would perhaps propose that experience in this context is kind of a fugue state, a possible solution to a digital problem: ambiguity as an antidote to information. With all the problems a suggestion of this sort poses, the problems are, at least, lived and explored together rather than imposed by technical standards of operation.

Bibliography

Behar, Katherine. *Bigger than You: Big Data and Obesity* (Goleta: punctum books, 2016). https://doi.org/10.21983/P3.0135.1.00.

Behar, Katherine, ed. *Object-Oriented Feminism* (Minneapolis: University of Minnesota Press, 2016).

Boetzkes. Amanda. "Ecologicity, Vision and the Neurological System," in Heather Davis and Etienne Turpin, eds., *Art in the Anthropocene: Encounters Among Politics, Aesthetics, Environments and Epistemologies* (Ann Arbor: Open Humanities Press, 2015).

cárdenas, micha. *The Transreal: Political Aesthetics of Crossing Realities* (New York: Atropos Press, 2011).

Castaneda, Carlos. *A Journey to Ixtlan: The Lessons of Don Juan* (New York: Simon and Schuster, 1972).

Chevillard, Éric. *On the Ceiling*, Jordan Stump, trans. (Lincoln: University of Nebraska Press, 2000).

Flusser, Vilém. *Towards a Philosophy of Photography* (London: Reaktion Books, 2000).

Kataoka, Serena. "Feeling with Your Eyes: Petroculture and *Ice Follies*," in Ted Hiebert, ed., *Plastic Blue Marble—Catalyst: Amanda Boetzkes* (Seattle: Noxious Sector Press, 2016).

Kroker, Arthur. *The Will to Technology and the Culture of Nihilism: Heidegger, Marx, Nietzsche* (Toronto: University of Toronto Press, 2015).

Laruelle, François. *The Non-Philosophy Project: Essays by François Laruelle* (Candor: Telos Press Publishing, 2012), 219.

Laruelle François. *Photo-Fiction: A Non-Standard Aesthetics* (Minneapolis: Univocal Publishing, 2012).

Virilio, Paul. *Politics of the Very Worst* (New York: Semiotext(e), 1999).

Virilio, Paul. *The Information Bomb*, Christ Turner, trans. (London: Verso, 2000).

Chapter 2
From Analogue Facialities to Digital Probeheads: Deterritorializing the Photographic Composites of Francis Galton and Nancy Burson

Colin Gardner
University of California, Santa Barbara

Abstract

This chapter applies Gilles Deleuze and Félix Guattari's twin axes of faciality – the white wall of *signifiance* and the black hole of subjectification – to the generic and racist analogue composite portraiture of Francis Galton at the turn of the twentieth century, compared to Nancy Burson's digital equivalent, which breaks down the signifiance/subjectification structure of the face through a deterritorializing and transforming probehead. Although the latter never fully escapes the dominance of the face in terms of structure and signification, it turns them against themselves as an affective embracing of different temporalities and races (Burson is most famous for her *Time* magazine cover that morphs Trump and Putin). Burson's achieves her greatest success in her 1990s androgynous *He/She* series, which are all the more powerful because we expect them to be digital composites (and therefore "artificial") when in fact they are purely indexical.

Keywords: Galton, Burson, Deleuze, Guattari, Faciality, Probehead, Eugenics.

***

Although modern audiences are extremely familiar with digitally constructed photographic composites of the human face through pioneering artists such as Nancy Burson (famous for her morphing of Donald Trump and Vladimir Putin on the July 30, 2018 cover of *Time* magazine), it's important to note that the biologist Sir Francis Galton (1822-1911), the cousin of Charles Darwin and founder of eugenics, made the first composites in analogue form as early as 1877. On the one hand, Burson's work defies easy classification of types and aims for a highly affective deterritorialization of facial stereotypes (what Gilles Deleuze and Félix Guattari call "Faciality," the twin axes of signifiance (white wall) and subjectification (black hole) whereby two very different (but mutually dependent) semiotic systems come together through the contours of the human face). In contrast, Galton deliberately created classifiable stereotypes by constructing composite pictures of the average criminal, the consumptive patient, and, on a more utopian level, aided by Darwin's son Leonard, a combination of officers and enlisted men from the British Royal Engineers (to prove the common excellence of the "English stock," regardless of class and rank). Although the latter composite was a little blurry, according to Galton, it had "an expression of considerable vigour, resolution, intelligence and frankness."[1] Galton's main objective was to isolate the physical characteristics of people thought to be carrying specific hereditary traits, whether it be crime or disease, such as psysthsis/tuberculosis (i.e. negative eugenics), all the better to isolate and marginalize such groups through selective breeding so that only superior strains – such as the military, the church, and the sciences – will prevail (positive eugenics), a case of nurture over nature.

On a more sinister level, he also created dystopic composites of what he considered to be inferior "races" such as diasporic Jews and the Irish (who were a major subject of racial discrimination at that time, where they were often seen as descendants of itinerant Iberian/African races), both of whom were proving extremely difficult to assimilate into Anglo-Saxon culture.[2] Not surprisingly, the latter works proved to be a springboard for anti-semitic studies in Germany, subsequently taken up by the Nazis. As Alexander Todorov points out,

> Galton's betterment of humanity involved the breeding of superhumans and restricting the breeding of "suboptimal humans." Spending the second half of his life promoting the "science" of eugenics, he was eventually successful. The first organized eugenics society was founded in Germany in 1905 and was called the Race Hygiene Society. Galton was the honorary president. Similar societies sprang up in the United Kingdom and the United States.[3]

Todorov stresses the impact of the German eugenicist, Hans F.K. Günther (1891-1968), also known as "Rassengünther" or "Race Günther" and "Rassenpapst" or "Race Pope") who followed Galton's logic to identify "superior" Nordic types. Significantly, his inaugural lecture as Professor of Social Anthropology at the

---

[1] Francis Galton, *Inquiries into Human Faculty and Its Development* (Everyman's Library edition, E.P. Dutton, 1928), 10.

[2] For illustrative examples of anti-Irish racism from the 13th Century to the 1980s see Liz Curtis, *Nothing But the Same Old Story: The Roots of Anti-Irish Racism* (Sásta, 1996).

[3] Alexander Todorov, *Face Value: The Irresistible Influence of First Impressions* (Princeton University Press, 2017), 26.

University of Jena in 1930 was attended by both Adolf Hitler and Hermann Göring.

Although Galton saw his "Jewish type" composites as his most successful – he was asked to construct them in 1883 by Joseph Jacobs, co-founder of The Jewish Historical Society, "who was attempting to demonstrate the existence of a relatively pure racial type of modern Jew, intact despite the Diaspora"[4] – Dr. Adolf Neubauer, an Oxford scholar of Hebrew scripts who wrote an anthropological book about Jews in China, "asserted that the Jews were so racially mixed they could not be distinguished from the other European races on physiological grounds. He also maintained that if there was one physical trait that might be said to distinguish Jews, it was something in their expression, but this was the result of environmental factors and had nothing to do with heredity."[5] As Neubauer argued, stressing social and environmental factors over the racial bias of hereditary eugenics:

> The bulk of the Jews, if dressed in European costume, would present few characteristics by which they could be recognized. The only strong similarity existing amongst the living Jews is that they often still bear, more or less, the stamp of oppressed individuals, a stamp which they cannot get rid of quickly, since there are many countries even now where they are treated like pariahs, and from whence they emigrate to more civilized lands.[6]

So, given the analog nature of Galton's composite experiments, how did he develop his method technologically in order to create physiognomic evidence that could hold up scientifically against such sociologically-based political readings? Initially, he took drawings and reduced them to the same scale so that he could trace them onto separate pieces of transparent paper. These were then superimposed onto each other and held up to the light so that they blended and overlapped. Obviously, this was an ephemeral process – nothing could be archived as scientific evidence – which led him to turn to photography. As Galton himself explained:

> I begin by collecting photographs of the persons with whom I propose to deal. They must be similar in attitude and size, but no exactness is necessary in either of these respects. Then, by a simple contrivance, I make two pinholes in each of them, to enable me to hang them up one

---

[4] Allan Sekula, "The Body and the Archive," *October* 39, no. 80 (Winter, 1986): 51-2.

[5] Anne Maxwell, *Picture Imperfect: Photography and Eugenics, 1870-1940* (Academic Press, 2010), 89.

[6] Dr. Adolf Neubauer, "Notes on the Race-Types of the Jews," *Journal of the Anthropological Institute* 15 (1885-86), 19.

> in front of the other, like a pack of cards, upon the same pair of pins, in such a way that the eyes of all the portraits shall be as nearly as possible super-imposed; in which case the remainder of the features will also be superimposed nearly enough. These pinholes correspond to what are technically known to printers as "register marks." They are easily made: A slip of brass or card has an aperture cut out of its middle, and threads are stretched from opposite sides making a cross. Two small holes are drilled in the plate, one on either side of the aperture. The slip of brass is laid on the portrait with the aperture over its face. It is turned about until one of the cross threads cuts the pupils of both the eyes, and it is further adjusted until the other thread divides the interval between the pupils in two equal parts. Then it is held firmly, and a prick is made through each of the holes. The portraits being thus arranged, a photographic camera is directed upon them.[7]

As in most of Galton's work, the entire exposure strategy is purely mathematical. If there are eight portraits in the pack, each is placed on the same sensitized plate and exposed for exactly ten seconds. "The sensitised plate will now have had its total exposure of eighty seconds; it is then developed, and the print taken from it is the generalised picture of which I speak. It is a composite of eight component portraits. Those of its outlines are sharpest and darkest that are common to the largest number of the components; the purely individual peculiarities leave little or no visible trace. The latter being necessarily disposed equally on both sides of the average, the outline of the composite is the average of all the components."[8]

The results are fairly consistent both facially and aesthetically. *The Composite Portrait of the Violent Criminal* (ca. 1880) resembles a fairly typical mug shot that you might see on a passport or driver's license, but all the facial detail has been bleached out, so that we are left with a white/dark contrast that accentuates the eyes, nostrils and lips. The same is true for the series of composites of *Psysthsis Patients from Bethlem Hospital* (1883), where one composite is formed from twelve individual portraits, another from six. The original portraits contain a broad cross-section of facial traits, including thick beards, ruffled hair, moustaches and deep-set eyes, while the two composites are pale, blurred and uniform in terms of facial homogeneity. As Anne Maxwell points out in her aptly titled book, *Picture Imperfect: Photography and Eugenics, 1870-1940*:

---

[7] Francis Galton, "Composite Portraits, Made by Combining Those of Many Different Persons Into a Single Resultant Figure," *The Journal of the Anthropological Institute of Great Britain and Ireland* 8, (1879), 133.

[8] Galton, "Composite Portraits," 134.

> As with the photographs of prisoners, Galton hoped to isolate the physical features that indicated mental illnesses so that the people who were prone to this condition might be singled out before they could reproduce, but the features of the patients were so irregular they could not be easily combined for averaging. Nor did things work smoothly with the composites of psysthsis patients. He was expecting the face in the composite to conform to the narrow ovoid type since this was the physiognomy anecdotally associated with the disease. However, not only was the resulting face much broader than he expected, but there was no noticeable difference between it and the one developed from a control group of people with other illnesses.[9]

Far from challenging Galton's belief that photographing physiognomies could help us diagnose hereditary diseases, he concluded instead that "a delicate narrow ovoid face may mean liability to other diseases not necessarily tubercular."[10]

This raises the importance of Deleuze and Guattari's notion of Faciality as a means of understanding Galton's eugenic reading of the facial composite. As we noted above, faciality is composed of two different axes: the white wall of signifiance and the black hole of subjectification. As Deleuze and Guattari argue in *A Thousand Plateaus*, "Signifiance is never without a white wall upon which it inscribes its signs and redundancies. Subjectification is never without a black hole in which it lodges its consciousness, passion, and redundancies. Since all semiotics are mixed and strata come at least in twos, it should come as no surprise that a very special mechanism is situated at their intersection. Oddly enough, it is a face: the *white wall/black hole* system. A broad face with white cheeks, a chalk face with eyes cut in for a black hole."[11] The key statement that reinforces Galton's mathematical and scientific position is that:

> Concrete faces cannot be assumed to come ready-made. They are engendered by an abstract machine of faciality, which produces them at the same time as it gives the signifier its white wall and subjectivity its black hole. Thus, the black hole/white wall system is, to begin with, not a face but the abstract machine that produces faces according to the

---

[9] Maxwell, *Picture Imperfect*, 86-7.

[10] Francis Galton and Dr. Mahommed, "An Inquiry into the Physiognomy of Phthisis by the Method of 'Composite Portraiture,'" *Guys Hospital Reports* 25, (1886), 490. Cited in Maxwell, *Picture Imperfect*, 87-8.

[11] Gilles Deleuze and Félix Guattari, *A Thousand Plateaus: Capitalism and Schizophrenia*, trans. Brian Massumi (University of Minnesota Press, 1987), 167.

> changeable combinations of its cogwheels. Do not expect the abstract machine to resemble what it produces, or will produce.[12]

Deleuze and Guattari also point out that while the head is included in the body as a whole (and is thus open to gestic communication of a social and political nature, à la Brecht's *gestus*, which is founded on citation and social relations), the face is completely separate, working as a kind of map that overcodes the body as a whole.

We see this in all aspects of modern life, in particular the dominant faciality of automobiles, where the headlights act as human eyes, and the grille stands in for the nose and mouth. Indeed, many of today's cars resemble the helmets of the stormtroopers from the *Star Wars* franchise, so that there is nothing more provoking than seeing that aggressive "face" in your rear-view mirror while you're being tailgated on the freeway (the mechanical face is literally "in your face"). However, faciality is also present in more subtle and benign ways, so that the curvature of the tops of trees in a landscape may resemble the crown of the human head, while, as Tom Conley argues, "the rotundity of a person's cheeks can resemble hillocks or mesas; the eyes might be reflective pools and ponds; the nostrils lairs and caves, and the ears at once quarries and cirques. Yet the landscape or face also looks at its spectators, calling their gaze into question or even psychically 'defacing' them."[13]

Galton pushed this one step forward because his focus on the composite face allows him to over-emphasise eugenics as beyond the human to the point that it becomes inhuman and indifferent to affective identification. As Deleuze and Guattari conclude, "It would be an error to proceed as though the face became inhuman only beyond a certain threshold: close-up, extreme magnification, recondite expression, etc. The inhuman in human beings: that is what the face is from the start. It is by nature a close-up, with its inanimate white surfaces, its shining black holes, its emptiness and boredom. Bunker-face."[14] In short, an excellent description of Galton's composites.

Given the analog nature of these composites, it's also worth exploring how they fit into Charles Sanders Peirce's triadic semiology of signs – the iconic, the indexical and the symbolic – so that we can draw a broader comparison when we discuss Burson's digital works. The iconic sign is any representational image that resembles a conceptual object or duplicates its properties, and, as Kaya

---

[12] Deleuze and Guattari, *A Thousand Plateaus*, 168.

[13] Tom Conley, "Faciality," in *The Deleuze Dictionary, Revised Edition*, Adrian Parr, ed., (Edinburgh University Press, 2005), 101.

[14] Deleuze and Guattari, *A Thousand Plateaus*, 170-1.

Silverman points out, "The most obvious icons are photographs, paintings, sculptures, and cinematic images, but algebraic equations and graphs are also iconic."[15] Galton's composites are also iconic both in terms of reproducing qualities of the different faces and also how they create a diagram of an overriding concept that unites them (such as criminality or disease). In *Camera Lucida*, Roland Barthes calls this general social and cultural context of the icon the "stadium" due to the shared cultural codes necessary for interpretation.[16] However, photography is most commonly seen as an indexical sign because it has a direct, causal and existential connection to its object that it directly captures onto the photosensitive surface (other indexical signs include the weathervane (wind + direction), a pointing hand (site specificity + direction) and a symptom (a rash, swollen glands or, in our case, the facial characteristics of Galton's psysthsis patients). Barthes sees the indexical as a form of "punctum" or prick that has a subjective impact on the viewer, generating the trace of a past moment or buried memory through the presence of a detail or partial object.[17]

Finally, the symbolic code has a completely arbitrary connection to the object, as in the case of all languages. However, like Galton's "faces," symbols are also governed by an overriding Law (in this case, the science of eugenics) which gives them a fixed, conceptual unity. As Peirce himself argues:

> A Symbol is a sign which refers to the Object that it denotes by virtue of a law, usually an association of general ideas, which operates to cause the Symbol to be interpreted as referring to that Object... Not only is it general itself, but the Object to which it refers is of a general nature. Now that which is general has its being in the instances which it will determine. There must, therefore, be existent instances of what the Symbol denotes, although we must here understand by "existent," existent in the possibly imaginary universe to which the Symbol refers.[18]

The scientific legitimacy of Galton's composites (and by extension Deleuze and Guattari's concept of Faciality) is purely arbitrary and symbolic, using the indexical characteristics of photography to make the sign seem less arbitrary, so that the generic (studium) and the specific (punctum) are in constant interaction, even if they are completely dissimilar and often have nothing in common. As Allan Sekula points out in his seminal essay, "The Body and the Archive," Galton's

---

[15] Kaya Silverman, *The Subject of Semiotics* (Oxford University Press, 1983), 19.

[16] Roland Barthes, *Camera Lucida: Reflections on Photography*, trans. Richard Howard (The Noonday Press, 1981), 27-8.

[17] Barthes, *Camera Lucida*, 43-5.

[18] Charles Sanders Peirce, *Collected Papers, Vols. I-VIII*, Charles Hartshorne and Paul Weiss, eds., (Harvard University Press, 1931), Vol. 11: 143-4.

entire project is about using the face to build an archive, in this case, one modelled on the military:

> In retrospect, the Galtonian composite can be seen as the collapsed version of the archive. In this blurred configuration, the archive attempts to exist as a potent single image, and the single image attempts to achieve the authority of the archive, of the general, abstract proposition. Galton was certainly a vociferous ideologue for the extension and elaboration of archival methods. He actively promoted familial self-surveillance for hereditarian purposes, calling for his readers to "obtain photographs and ordinary measurements periodically of themselves and their children, making it a family custom to do so." [...] His model here was the British Admiralty's voluminous registry of sailors. Here again, eugenics modeled itself on the military.[19]

As we will see when we discuss Burson, we must escape the face (and, by extension, the archive) in order to transform the abstract qualities of the symbol back into the human as a form of sensate becoming.

However, to his credit, when Galton worked outside the limited parameters of faciality, his composite experiments were far more avant-garde and, some might argue, a precursor to contemporary digital representations of machinic movement. In contrast to Eadward Muybridge's famous 1878 work, *The Horse in Motion*, which showed Leland Stanford's horse, Sallie Gardner galloping in twelve photographic freeze frames on the Palo Alto racetrack (illustrating how the camera could show us what the human eye could only imagine), Galton's 1882 equivalent followed Ētienne-Jules Marey's lead (typified by his 1886 photograph, *Cheval blanc monté*) in showing the horse's legs and hooves in a simultaneous blurring of movement, an effect exacerbated by the image's TV-like horizontal analog raster lines. This is a direct forerunner to Marcel Duchamp's painting, *Nude Descending a Staircase, No 2* (1912) and also, more importantly, Giacomo Balla's *Dynamism of a Dog on a Leash* (1912), which features the blurred movement of a dachshund alongside the feet of its female owner. The latter was a major example of the Italian Futurists' interest in the merging of the machinic with the human through a common dynamism, which, if we ignore the movement's later connection to Italian fascism, is a key component of Deleuze and Guattari's interest in what they call The Body Without Organs or BwO, particularly its role in constructing "the conjunctive synthesis" or "Celibate Machine" which reconciles the difference between repulsion and attraction (as opposed to Galton who spent his entire career trying to isolate them eugenically). Using Duchamp's *Large Glass* (aka. *The Bride Stripped Bare*

[19] Sekula, "The Body and the Archive," 54.

by Her Bachelors, Even, 1915-23), as their key example for the machinic generation of pure intensities, Deleuze and Guattari argue that "A genuine consummation is achieved by the new machine, a pleasure that can rightly be called autoerotic, or rather automatic: the nuptial celebration of a new alliance, a new birth, a radiant ecstasy, as though the eroticism of the machine liberated other unlimited forces."[20] Which, as it turns out is an excellent summary of Burson's artistic objectives, as well as the counter to the dominance of Faciality, "the probe-head."

After studying painting for two years at Colorado Women's College, in 1968, Burson moved to New York, where she was immediately inspired by the interactive MOMA exhibition, *The machine, as seen at the end of the mechanical age.* She decided that she wanted to construct a form of "aging booth" or *Age Machine* – "a simple machine where you could push a few buttons and automatically see yourself older."[21] She then visited E.A.T. (Experiments in Art and Technology), a New York-based non-profit where she met Carl Machover, a computer graphics expert, hoping to get the project off the ground, but unfortunately, Machover explained to her that computer technology still had a long way to go to meet her needs. It turned out to be an eight-year delay, but in 1976, Burson finally began a groundbreaking collaboration with Tom Schneider and David Petty, leading engineers in the Architecture Machine Group at M.I.T. As Lynn M. Herbert explains, two new developments in computer technology made her aging booth concept realizable:

> An analog-to-digital converter could change a video signal into a computer signal, and a frame buffer allowed a computer to store and manipulate graphic images. Together, the artist and the engineers developed a program that allows a computer to interact with an image of a face. Portraits are scanned by a television camera and digitally encoded into a computer. In 1981, Burson received a U.S. Patent for "The Method and Apparatus for Producing an Image of a Person's Face at a Different Age."[22]

In collaboration with Cambridge-based computer scientists Richard Carling and David Kramlich, Burson refined the system technologically and, moving beyond the confines of the gallery and museum-based art world, their first aging portraits – what Burson calls "morphing" – were published in *People*

---

[20] Gilles Deleuze and Félix Guattari, *Anti-Oedipus: Capitalism and Schizophrenia*, trans. Robert Hurley, Mark Seem, and Helen R. Lane (University of Minnesota Press, 1983), 18.

[21] "Transcript of France Morin's interview with Burson for an exhibition of *The Age Machine* at The New Museum of Contemporary Art, New York, January 1992, cited in Lynn M. Herbert, "Faces: Nancy Burson," in *Faces: Nancy Burson* (Contemporary Arts Museum, Houston, 1992), 8.

[22] Herbert, "Faces: Nancy Burson," 8.

magazine in 1982, featuring images of Lady Diana, Princess of Wales, actors Timothy Hutton, Brooke Shields and John Travolta, all "aged" way before their time. Unlike Galton's eugenically-based archival composites, based on inbreeding, Burson was far more interested in questions such as "what if...?" thereby creating new identities (or criteria for new identities) that people had never considered possible.[23] Thus, for example, in 1982 she created two different beauty composites, one featuring Bette Davis, Audrey Hepburn, Grace Kelly, Sophia Loren and Marilyn Monroe, the other Jane Fonda, Jacqueline Bisset, Diane Keaton, Brooke Shields, and Meryl Streep, so that we became immediately aware of the shifting nature of Hollywood's idea of the female film star persona from the 1950s to the 1980s. The result is not unlike Cindy Sherman's non-composite photographic self portraits – the *Untitled Film Stills* series – that evoke typical (if not clichéd) performative personae.[24]

However, it's important to note that Burson has a far greater concern with social consciousness when it comes to her composites, typified by her raising a key question: What would a missing child look like several years after their disappearance? As Lynn Herbert rightly argues:

> "Updating" missing children is an example of how Burson's work lives multiple lives inside and outside the art world. Beginning in 1982, parents of missing children began to contact Burson. They had seen her "aged" portraits exhibited, published and on television. Could she provide an updated composite of their missing child? It was not a project Burson had foreseen, but she set about to help these parents. Her computer program for aging only worked on persons 18 years of age or older, an age when the facial structure is fully formed. She went back to the

---

[23] Interestingly, at that early point in her career, Burson had never heard of Galton, and he was not an influence on her early experiments. As she explained in an interview with Lynn Gumpert and Terrie Sultan, 'I never heard of Francis Galton until I was deep in the heart of composite country. I heard about him at the first computer graphics conference that I attended, where I showed the videotape made at MIT of the three people aging. When I started working on my first book, *Composites*, I went to the public library to do some research. I was the first person in many, many decades to check out Galton's book. It was pretty exciting. I came up from the bowels of the library; it had not been touched in decades.' Lynn Gumpert and Terrie Sultan, "Interview with Nancy Burson," in *Seeing and Believing: The Art of Nancy Burson* (Twin Palms Publishers, 2002), 151.

[24] Burson also created two composites of male film stars from different eras and used the second grouping, composed of Richard Gere, Christopher Reeve, Mel Gibson, Warren Beatty and Robert Redford in combination with an ape to transcend evolutionary time. The result – *Evolution II* (1984) – looks alarmingly close to a composite, prehistoric portrait of George W. Bush and Tony Blair.

> drawing table and discovered that she could "age" younger children by compositing a photograph of the child with a photograph of the sibling or parent with the closest resemblance.[25]

To give some idea of the social import of Burson and Kramlich's project, their program has been used extensively by both the F.B.I. and the National Center for Missing and Exploited Children to track down both children and adults. Perhaps the most well-known example is that of Etan Patz, who was "updated" from age six to thirteen in 1984. Born in October 1972, Etan disappeared on May 25th, 1979, at the age of six, on his way to catching the school bus in Lower Manhattan. Burson was particularly pleased with her composite, for "As in all composites, the better the quality of the photographs used, the better the composite. One reason why the 'Etan Patz Update' is the most successful is the excellent quality of the original photos used. And for a good reason – Etan's father is a commercial photographer."[26] While Burson's high-grade composite gave everyone hope insofar as it gave the police concrete visual evidence to go on, it is now believed that Etan was abducted and murdered on the day of his disappearance. Again, unlike Galton's detached, "scientific" composites, Burson's contribution to the case is full of empathy and affect. Although she herself kept a deliberate, anonymous distance from each case, the audience – whether artistic or familial – is always directly involved in the vicissitudes of real life. Dana Friis-Hansen summarizes it well when she notes that, "Lost without a trace, the children return as adolescents through ghostly silent pictures. These images should remind us that while growing older can be registered in the changes on one's face, and even coded into a computer, true growing up is felt through the changes experienced in one's life."[27]

Burson's attachment to the social and political becomes even more profound with her 1982 composite, entitled *Three Assassins*, which combines portraits of Lee Harvey Oswald, Sirhan Sirhan and James Earl Ray. As Burson later explained, "This was one of our first composites, done at a time when we hadn't perfected the software for cleaning up an image. I had expected a portrait of evil and ended up with the boy next door."[28] Herbert rightly argues that "in Burson's resulting sweet face, Galton would be hard pressed to find any innate criminal

---

[25] Herbert, "Faces: Nancy Burson," 9-10.

[26] Nancy Burson, Richard Carling, and David Kramlich, "Notes on Some Composites by Nancy Burson," in *Composites* (Beech Tree Books, William Morrow, 1986), 94-5.

[27] Dana Friis-Hansen, "Nancy Burson's Strange Mirrors," in *Nancy Burson: "The Age Machine" and Composite Portraits* (MIT List Visual Arts Center, 1990), 5.

[28] Burson, Carling, and Kramlich, "Notes on Some Composites by Nancy Burson," 94.

tendencies"[29] but on the other hand, as in all of her works, Burson urges us to question the validity of the line-up in the first place: are these real assassins or patsies? (a political question that Galton would probably never have posed).

Burson also tackled mainstream politics with *Warheads I, II, III, & IV* (1982) and *Big Brother* (1983). In the former, as Burson explains, "Each composite is weighted to the number of nuclear warheads deployable by each country at the time each leader served (Source: Council for a Nuclear Weapon Freeze, Cambridge, MA)."[30] *Warhead I*, for example, features combinations of Ronald Reagan, 9,500 warheads (55%); Leonid Brezhnev, 7,700 (45%); Margaret Thatcher, 64 (less than 1%); François Mitterand, 64 (less than 1%) and Deng Xiaoping, 64 (less than 1%). Interestingly, the dominant face that emerges from the composite is that of Brezhnev, even though he has fewer warheads than Reagan, suggesting that bushy eyebrows, haggard features and squinty eyes become a dominant facial trope (running counter to Galton's more "washed-out" composites that accentuate pale cheeks and foreheads). We see a similar effect in *Big Brother.* According to Burson, "CBS-YV commissioned this for a Walter Cronkite special entitled *1984 Revisited*, in which this photograph served as the visual backdrop for a discussion of the realization of George Orwell's vision of 1984. At the end of the program, Cronkite said, 'We put these posters up – we can take them down… this time.' Then he ripped it off the wall, tore it up, and threw it in a gutter."[31] The composite is composed of Josef Stalin, Benito Mussolini, Mao Zedong, Adolf Hitler and Ruhollah Khomeini, but defies all expectations in terms of dictatorial facial stereotypes. One would expect Hitler's face to be far more prominent, but of course, his Chaplinesque "toothbrush" moustache is completely transgressed and superseded by Stalin's black handlebar variation. Also, the latter's nose and eyebrows merge perfectly with those of Khomeini to create a far more non-Western 'look.'[32]

So far, we have explored how Burson's use of digital technology creates a far more affective, non-archival approach to composite portraits. But what happens if we take a common subject – such as androgyny – and compare the results from the early 1980s when Burson computerized the faces, to her late 1990s analog, non-composite equivalents – the *He/She* series? Does one revolutionize the other in terms of our reading and reception of difference, and how do they

---

29 Herbert, "Faces: Nancy Burson," 8.

30 Burson, Carling, and Kramlich, "Notes on Some Composites by Nancy Burson," 94.

31 Burson, Carling, and Kramlich, "Notes on Some Composites by Nancy Burson," 94.

32 Interestingly, given Burson's later merging of the faces of Donald Trump and Vladimir Putin for *Time* magazine, one of the most prominent American flaunters of the toothbrush moustache (from 1940 to 1950) was Trump's father, the real estate developer, Fred Trump (1905-99). A Burson composite of Fred and Adolf would be fascinating!

undermine Faciality? Let's start with her 1982 composite, *Androgyny*, from 1982. As Burson herself explained:

> For this amalgam of six men and six women to be unbiased, I felt it was important to mix twelve totally dissimilar faces. I had no expectations but was nonetheless surprised at the domination by the female characteristics, though I would come to realize later that the female dominates whenever male and female are mixed. The only male characteristic here is the mouth; cover the mouth and you find a wholly feminine face. Perhaps the reason for this lies partly in the process – when you "average" faces together, rougher features are softened, which makes them read as more feminine. However, I prefer to think that in a larger sense, we are all conceived as females. To most of us, an infant, at first glance, seems more female than male. The feminine is primary. We evolve from it.[33]

Although this may seem like Burson is arguing that the female is the basic unit of the face (i.e. she seems to be falling into Galton's eugenic trap of the law of the symbolic code), by separating out the gendered quality of the composite's features we see that the mouth is stubbornly male, so that we are encouraged to view the image indexically for contradictory traits. Even if the female face dominates, the male is still conceptually present.

Fifteen years later, when Burson had embraced analog, non-composite portraiture, she developed her *He/She* series of Polaroids and Iris prints. As she explains, "These are people who are on the edge of male and female – they emphasize this, and they want to make it hard for people to differentiate. Whether or not that comes from an internal conflict is another issue. What I found is that there is a very playful aspect to their androgyny. The fact that they appeal to both sexes is very important to them."[34] Of course, the key element here is that because this is a series of works by Nancy Burson, we expect the faces to be digital composites, artificially blurring the distinction between male and female through advanced technology and morphing. When we discover that they are actually analog non-composites, the whole process becomes an unresolvable guessing game with no right or wrong answer. "In my studio, no one has ever gotten them all right," admits Burson. "Interestingly, according to studies, most women are attracted to feminine men. I've also found that it's harder for men to look feminine than for females to look masculine."[35] She also

33 Burson, Carling, and Kramlich, "Notes on Some Composites by Nancy Burson," 94.

34 Gumpert and Sultan, "Interview with Nancy Burson," 149.

35 Gumpert and Sultan, "Interview with Nancy Burson," 149.

notes that those who posed for her didn't have to make their gendered image as confusing as it turned out – the results were an intrinsic part of their self-image and self-acceptance: "That is the most difficult task we face – to love ourselves. When we do, then difference doesn't matter."[36]

This effect is reinforced by the faces being photographed in close-up and bathed in an orange *chiaroscuro* reminiscent of Rembrandt's painted portraits, so that, by merging the iconic semiotics of art historical representation and contemporary issues of gendered identity, we are forced to think outside the conventional boundaries of indexical and symbolic difference. Also, unlike in *Androgyny,* the mouth is no longer dominant as a specifically male trait – all facial features are equally gendered. As Michael Sand rightly argues, "Floating in black backgrounds, the faces demonstrate that the relativity of beauty and gender are interrelated by suggesting alternate versions of ourselves."[37] More importantly, Burson revolutionizes analogue photography by using digital composite discourse to liberate its potential beyond the indexical and Galton's use of the medium for eugenics. The analog has now moved far beyond Faciality and has instead become a powerful probe-head. So what exactly is a probe-head?

Inspired by the work of Francis Bacon, Deleuze and Guattari make the provocative argument in *A Thousand Plateaus* that:

> if human beings have a destiny, it is rather to escape the face, to dismantle the face and facializations, to become imperceptible, to become clandestine, not by returning to animality, nor even by returning to the head, but by quite spiritual and special becomings-animal, by strange true becomings that get past the wall and get out of the black holes, that make faciality traits themselves finally elude the organization of the face – freckles dashing toward the horizon, hair carried off by the wind, eyes you traverse instead of seeing yourself in or gazing into in those glum face-to-face encounters between signifying subjectivities.[38]

The most effective counters to Faciality are the above-mentioned probe-heads (*têtes chercheuses,* guidance devices) that dismantle the face's strata by breaking through the walls of significance and pouring out of the holes of subjectivity (i.e., the white wall/black hole machine) as deterritorializing rhizomes or lines of flight, much like Burson's fluid becoming-other in the

---

[36] Gumpert and Sultan, "Interview with Nancy Burson," 150.

[37] Michael L. Sand, "Nancy Burson and the Art of Seeing," in *Seeing and Believing: The Art of Nancy Burson* (Twin Palms Publishers, 2002), 19.

[38] Deleuze and Guattari, *A Thousand Plateaus,* 171.

He/She series. Simon O'Sullivan sums it up well when he argues that probe-heads constitute

> lines of escape from the face and from faciality. Crucially, they are not a return to some kind of primitive pre-faciality. They are, in fact, an escape that takes place from within the terrain of the face, a kind of stammering from within. We might add that probe-heads need not necessarily be pictures of heads but rather any device that disrupts faciality, for the latter applies not just to heads but to most of the mechanisms that produce significance and subjectivity, from faces and landscapes within painting to facialisation and landscapification within the world.[39]

In other words, probe-heads use the innate structure of faciality – the white wall/black hole machine – against itself so that, for example, when we look at a landscape we perceive a multiplicity of potential significations, not unlike Cézanne's paintings of Mont Sainte-Victoire, which are far more angular and geometric, almost Cubistic, compared to their neo-realist, Impressionistic equivalent.

In conclusion, it is worth noting that Burson's role as a visual artist allows her to harness the deconstructive power of the probe-head – not unlike the puncturing effect of Barthes's photographic punctum – to break down the parameters of faciality far beyond the limitations of both indexicality and the symbolic. More importantly, this is a collective rather than a personal exercise, for as Burson states in her interview with Lynn Gumpert and Terrie Sultan, "As far as I'm concerned, my service as a human is imaging the paradigm shift from 'I-ness' to 'We-ness.' In the eyes of the universe, we are all parts of the whole."[40] However, we should also acknowledge the fact that this "whole" is historical as well as spatial and communal, for Galton's analogic composites have laid a key foundation for understanding and appreciating the importance of Burson's practice (the *He/She* series wouldn't be nearly as effective without his eugenically-based precursors). In this sense, art transcends historical time and place, for as O'Sullivan rightly argues, it also looks to the future as a form of practice-to-come: "Contemporary art can operate on a cusp between the present and the future. It is 'made' in the present, out of the materials at hand, but its 'content' often calls for a subjectivity to come. This stuttering and stammering of existent materials and languages, this deterritorialization of existing regimes of signs,

39 Simon O'Sullivan, *Art Encounters Deleuze and Guattari: Thought Beyond Representation* (Palgrave Macmillan, 2006), 60.

40 Gumpert and Sultan, "Interview with Nancy Burson," 154.

constitutes the ethico-aesthetic function of art."[41] In other words, the probe-head guarantees difference as the foundation for de-facializing the face, but also acts as the agent for all future artistic creation, not only that of Nancy Burson but all depictions of the human face in all media. The face may never go away, but it is always open to endless change.

Bibliography

Barthes, Roland, *Camera Lucida: Reflections on Photography,* trans. Richard Howard (The Noonday Press, 1981).

Burson, Nancy, Richard Carling, and David Kramlich, "Notes on Some Composites by Nancy Burson," in *Composites* (Beech Tree Books, William Morrow, 1986).

Conley, Tom, "Faciality," in *The Deleuze Dictionary, Revised Edition,* Adrian Parr, ed., (Edinburgh University Press, 2005).

Curtis, Liz, *Nothing But the Same Old Story: The Roots of Anti-Irish Racism* (Sásta, 1996).

Deleuze, Gilles and Félix Guattari, *Anti-Oedipus: Capitalism and Schizophrenia,* trans. Robert Hurley, Mark Seem, and Helen R. Lane (University of Minnesota Press, 1983).

Deleuze, Gilles and Félix Guattari, *A Thousand Plateaus: Capitalism and Schizophrenia,* trans. Brian Massumi (University of Minnesota Press, 1987).

Friis-Hansen, Dana, "Nancy Burson's Strange Mirrors," in *Nancy Burson: "The Age Machine" and Composite Portraits* (MIT List Visual Arts Center, 1990).

Galton, Francis, "Composite Portraits, Made by Combining Those of Many Different Persons Into a Single Resultant Figure," *The Journal of the Anthropological Institute of Great Britain and Ireland* 8, (1879).

Galton, Francis, *Inquiries into Human Faculty and Its Development* (Everyman's Library edition, E.P. Dutton, 1928).

Galton, Francis and Dr. Mahommed, "An Inquiry into the Physiognomy of Phthisis by the Method of 'Composite Portraiture,'" *Guys Hospital Reports* 25, (1886), 490. Cited in Maxwell, *Picture Imperfect,* 87-8.

Gumpert, Lynn and Terrie Sultan, "Interview with Nancy Burson," in *Seeing and Believing: The Art of Nancy Burson* (Twin Palms Publishers, 2002), 151.

Herbert, Lynn M. "Faces: Nancy Burson," in *Faces: Nancy Burson* Museum, Houston, 1992).

Maxwell, Anne, *Picture Imperfect: Photography and Eugenics, 1870-1940* (Academic Press, 2010).

Neubauer, Dr. Adolf, "Notes on the Race-Types of the Jews," *Journal of the Anthropological Institute* 15 (1885-86).

O'Sullivan, Simon, *Art Encounters Deleuze and Guattari: Thought Beyond Representation* (Palgrave Macmillan, 2006).

[41] Simon O'Sullivan, *On the Production of Subjectivity: Five Diagrams of the Finite-Infinite Relation* (Palgrave Macmillan, 2012), 200.

O'Sullivan, Simon, *On the Production of Subjectivity: Five Diagrams of the Finite-Infinite Relation* (Palgrave Macmillan, 2012).

Peirce, Charles Sanders, *Collected Papers, Vols. I-VIII,* Charles Hartshorne and Paul Weiss, eds. (Harvard University Press, 1931), Vol. 11.

Sand, Michael L, "Nancy Burson and the Art of Seeing," in *Seeing and Believing: The Art of Nancy Burson* (Twin Palms Publishers, 2002),

Sekula, Allan, "The Body and the Archive," *October* 39, no. 80 (Winter, 1986): 51-2. https://doi.org/10.2307/778312.

Todorov, Alexander, *Face Value: The Irresistible Influence of First Impressions* (Princeton University Press, 2017).

Silverman, Kaya, *The Subject of Semiotics* (Oxford University Press, 1983).

Chapter 3
Bodies of Work: Ferrania Film's Analog Legacies in the Anthropocene

Elena Past
Wayne State University

Abstract

"Bodies of Work" investigates the analog legacies of Italian film stock manufacturer Ferrania by situating the factory's history in the context of the Anthropocene, a time of unfolding environmental catastrophe. Drawing on scholarship in labor history, environmental history, and new materialisms, as well as archival research about the Ferrania factory, my essay proposes that the *filmiste*, the factory's women workers, embody the apparent contradictions of what I call the analog Anthropocene. In the analog Anthropocene, analog film, which helped to envision, record, and archive the history of the twentieth century, can be understood both as a resource-intensive, potentially hazardous industrial product, and as an affective suture that binds together the histories and memories of the workers who made it.

Keywords: Ferrania Film, Anthropocene, Film Stock, Environmental Humanities, Sabrina Gschwandtner.

***

The Ferrania film factory, located in a rural mountain valley in the province of Savona in northwestern Italy, closed its doors in the early 2000s when digital photography definitively eclipsed analog. Today, the sprawling factory landscape is dominated by weeds, broken glass, and piles of rubble. Industrial ruin is one of the legacies of analog film, as the evocative photographs by Robert Burley about the disappearance and destruction of former film factories and labs have demonstrated.[1] But ruin is not the only legacy of analog film, or of Ferrania.

This chapter investigates Ferrania's analog legacies in the Anthropocene, a time of unfolding environmental catastrophe. My hypothesis is that, in dialogue, analog film and the Anthropocene help reveal the toxic inheritance

---

1 See Robert Burley, *The Disappearance of Darkness*, https://robertburley.com/_rb/disappearance-of-darkness/.

of the industrial twentieth century, while also offering affective affordances for moving forward in the twenty-first. I consider how analog film, the iconic "eye" of the twentieth century, can be understood simultaneously as a resource-intensive, potentially hazardous industrial product, and as an affective suture that binds together the histories and memories of the workers who made it. Working at the intersection of labor history, environmental history, and new materialisms, my essay suggests that the *filmiste*, the women workers in the Ferrania factory, embody the apparent contradictions of what I call the analog Anthropocene.

Using an eclectic variety of archival materials that include industrial documentaries and investigative reports, oral histories and conversations on Facebook, I consider the environmental impacts of making film, where by environment I primarily refer to the intimate sphere of the human bodies of the *filmiste*. Drawing on material ecocritical theory and labor ecology, I observe the bodily experiences of women who worked at Ferrania in the 1950s and 60s, in the postwar moment scientists have dubbed the Great Acceleration, when the earth witnessed a surge in various human-driven trends (population growth, energy use, ocean acidification, among other things), and when the factory was experiencing its golden age.[2] Laboring in a photochemical factory meant that the *filmiste* worked in the service of the future exposure of film stock to the light, and were in the process exposed to a host of chemical and mechanical dangers. Then, I follow stories of *filmiste* who, decades later, were remembered on YouTube and Facebook, appearing in the digital universe. These stories reveal how repetitive factory labor in many cases constituted a space of affective intimacy, an experience that endured long after jobs concluded. The interwoven threads of these stories, written across analog and digital media, create a patchwork view of analog's legacies in the twenty-first century: multiple generations of bodies, materials, and labor woven into the fabric of memory and matter.

The Anthropocene is Analog

Although, according to a recent vote, the name of our contemporary geological epoch will not be changed from "Holocene" to "Anthropocene" by stratigraphers, the Anthropocene has captured a significant place in the scientific and cultural imaginations. What does it mean to revisit *analog* with the *Anthropocene* front of mind? This association is significant because the Anthropocene, an era in which human impacts have begun to profoundly impact major earth systems

[2] Will Steffen, Wendy Broadgate, Lisa Deutsch, Owen Gaffney, and Cornelia Ludwig, "The Trajectory of the Anthropocene: The Great Acceleration," *The Anthropocene Review* 2, no. 1 (2015): 81-98, https://doi.org/10.1177/2053019614564785.

(the biosphere, atmosphere, lithosphere, and the oceans), is analog.[3] Analog is an imprint of one element of the world – sound, light, text – on another. The Anthropocene is the imprint of various worldly elements – pesticides, fertilizers, radionuclides, plastiglomerates – on others. Both the Anthropocene and analog are defined by material traces, and they both describe a practice that might be thought of as archival, since they are repositories of time, knowledge, and matter.

Analog is a form of relationality based on continuity: a continuous signal recorded as such in (or on) a new format, like the ridges and grooves on a vinyl record, or the assemblages of silver salts in a thin film of gelatin on a strip of celluloid, or the rain of nuclear fallout on earthly surfaces across the globe. The ridges and grooves, silver salts, and radionuclides have a direct relationship to the sounds or figures or events they record, but are not the same. Scholars have convincingly objected that the "Anthropos" of the Anthropocene does not represent the labors and trials of all humans everywhere, but the analog Anthropocene holds these contradictions. Analog is reproducible, but every copy, imprinted on a unique carrier with its own independent history of circulation and its own timeline, is itself unique.

Most significantly for the histories that unfold in the pages ahead, analog is also touch. It forms via what Giuliana Bruno calls the haptic, or "the reciprocal con-*tact* between us and our surroundings,"[4] or between surroundings and surroundings. Touch is a basic element of existence on a planet governed by gravity and surrounded by an atmosphere. Touch is always reciprocal: to touch is to be touched. It is also always processual: all parties are changed by way of contact, for better or for worse. Erin Manning writes in *Politics of Touch:* "Bodies are connected through intensities of composition that in turn produce new bodies."[5] The bodies of which she writes are mostly human bodies, the contact mostly skin to skin, but the reciprocity of which she writes is proper to all bodies in all of their shifting, telescoping complexities: film to finger, air to film, factory to valley, archivist to archive, etc.

---

[3] After years of work, the Anthropocene Working Group, a subcommittee of the International Commission on Stratigraphy, voted in favor of declaring the Anthropocene a formal chrono-stratigraphic unit in May of 2019. As I was writing this essay, the *New York Times* reported that geologists from the broader International Commission declined to affirm a change to the geological timescale. Raymond Zhong, "Geologists Make It Official: We're Not in an 'Anthropocene' Epoch," *New York Times*, March 20, 2024, https://www.nytimes.com/2024/03/20/climate/anthropocene-vote-upheld.html.

[4] Giuliana Bruno, *Surface: Matters of Aesthetics, Materiality, and Media* (Chicago: University of Chicago Press, 2014), 144.

[5] Erin Manning, *Politics of Touch: Sense, Movement, Sovereignty* (Minneapolis: University of Minnesota Press, 2006), xvi.

Why are these qualities of analog important in understanding the Anthropocene, the contemporary state of climate catastrophe? Prominent thinking and communication about climate change, as Stacy Alaimo has powerfully argued, too often relies on "meta-science" and what she calls "an epistemological stance of hyperseparation" that detaches itself from "a sense of exposure."[6] The imagery and language of technoscience, which "promises to deliver us utterly disembodied, transcendent, and complete knowledge,"[7] risks covering over the very particular vulnerabilities born in haptic exchange, in the direct contact that links all kinds of bodies on the planet. By instead focusing on material imprint and haptic exchange, the analog Anthropocene stubbornly "refuses to disavow our immersion within the material world."[8]

The Anthropocene is analog. As an imprint of world on world, and as a continuous (but not uniform) phenomenon across time and space, the Anthropocene is necessarily local and perceptible, with distinctive coordinates at each point in time and space. It can best be understood in the context of local histories and with the help of vibrant materials, like those manufactured at the Ferrania film factory in Liguria's Val Bormida.

Locating Ferrania Film

Starting in 1923, for much of the twentieth century and into the twenty-first, the Ferrania factory made photosensitive analog film stock for movies, photography, and X-rays. In the 1960s, the factory was acquired by Kodak competitor and American multinational giant Minnesota Mining and Manufacturing (3M) and began manufacturing products under a myriad of different brand names. In the early 2000s, with the definitive rise of digital, the factory was shuttered (although a Kickstarter campaign a few years later helped relaunch a tiny production facility in Ferrania's former research lab).[9]

Why does Ferrania matter in the history of cinema? From the 1930s through the 1960s, thousands of Italian-style comedies, industrial documentaries, newsreels, and auteurist masterpieces were shot on Ferrania negatives. Roberto Rossellini used Ferrania for his first feature-length film, *La nave bianca* (*The White Ship*, 1941). Federico Fellini's *Le notti di Cabiria* (*Nights of Cabiria*, 1956) was shot on Ferrania Pancro C7. All of Pier Paolo Pasolini's black and white

---

[6] Stacy Alaimo, *Exposed: Environmental Politics and Pleasures in Posthuman Times* (Minneapolis: University of Minnesota Press, 2016), 99, 102.

[7] Alaimo, *Exposed*, 101.

[8] Alaimo, *Exposed*, 102.

[9] The new, "artisan-industrial" scale Ferrania factory is producing panchromatic and orthochromatic film for still photography. For more information, see the Ferrania Film website, https://www.filmferrania.com/.

films, from *Accattone* (1961) to *Uccellacci e uccellini* (*The Hawks and the Sparrows*, 1966), were shot on Ferrania P30. Vittorio DeSica's 1960 film *La ciociara* (*Two Women*), for which Sophia Loren won a coveted Academy Award, used Ferrania P30. Some of the twentieth century's most iconic faces were visible on the surfaces of Ferrania's contrasty black-and-white film stock: Giulietta Masina's impish, playful grin as Cabiria; Anna Magnani's intense gaze as Mamma Roma; Sophia Loren's defiant glamor as Cesira in *Two Women*.

To convince film scholars and cinephiles of the relevance of studying film stock, a list of illustrious actors and directors conveys the cultural and aesthetic impact of this brand of photosensitive material. These are the names and faces that grace the headlines, the film posters, the title screens: the figures already canonized in the halls of film studies, the guarantors of Ferrania's cultural capital, and the most visible elements of the factory's analog legacy. But—to use a charged metaphor in an era of melting polar icecaps—this is only the tip of the iceberg, where the material-environmental histories of cinema are concerned. The history of photosensitive film stock has a lot to tell us about the unfolding of the twentieth and twenty-first centuries in environmental terms, as well.

Photosensitive film stock was the medium that carried countless stories—dramatic, comedic, epic, intimate—to eager audiences, and much ink has been spilled to analyze the images that crossed the silver screen. The significance of film stock's material-industrial history, however, is less prominent (although in the last decade or so this landscape of inquiry has begun to shift).[10] Film studies has often historically articulated its interests within the "dematerializing networks that cross through academic theory, popular culture, contemporary discourse, and everyday practice," a zone of inquiry in which, as Alaimo observes, nonhuman natures get a "rather shabby theoretical and rhetorical treatment."[11] The production of photosensitive analog film stock, a fascinating element of the material backstory of film production, invites readings of cinema's "emergent, ultimately unmappable landscapes of interacting biological, climatic, economic, and political forces."[12] The bodies of factory line workers and the complex chemical components of the photosensitive material that they collaborated in making offer evidence of the "trans-corporeal" interconnections that underpin cinema: a means to recognize film not as a mere projection of (human) desire

[10] See, for example, Alice Lovejoy, "Celluloid Geopolitics: Film Stock and the War Economy, 1939-47," *Screen* 60, no. 2 (2019): 224-241; and Pansy Duncan, "Towards a Natural History of Film Form: Silver Salts and the Aesthetics of Early Studio-Era Hollywood Cinema," *Screen* 64, no. 3 (2022): 411-426.

[11] Stacy Alaimo, *Bodily Natures: Science, Environment, and the Material Self* (Bloomington: Indiana University Press, 2010), 1-2.

[12] Alaimo, *Bodily Natures*, 2.

but also an integral part of a multispecies history. These less documented analog legacies, which originate in animals (gelatin), vegetables (cellulose), minerals (silver), and chemical labs, and later sediment in soils and bodies or off-gas into atmospheres, situate film history in the "narrative stratigraphy" of the Anthropocene.

The environmental impacts of film manufacture are the subject of my broader Ferrania project, but here my interest is focused on factory laborers. In writing of narrative stratigraphies in postwar Italy, Serenella Iovino observes that "the Anthropocene can [...] be read also in the way it stratifies in human bodies, in how these bodies respond to these challenges."[13] I'll now turn to investigate how traces of analog film manufacture registered within the bodies of Ferrania's workforce.

Life, Exposed: Bodies and *Work in Ferrania*

To argue that the Anthropocene is analog is to correct elements of the historical record that overestimated human mastery over matter – a transcendent narrative that has often distanced itself from histories of labor, and one that Alaimo recognizes as "hypermasculine" and apt to "detach [itself] from a sense of exposure."[14] Against this detachment, the historian Stefania Barca explains that "placing the workplace center stage in our narratives and understanding it as an ecological system" helps unfold the complexities of environmental history in the twentieth century. Centering the workplace in environmental history, in turn, allows the bodies of the workers to "fully enter environmental history narratives as meta-texts where the political ecology of industrial societies had been written."[15]

In the case of the Ferrania factory, the convergence of materials and lives on the production line – not just human bodies, but nonhuman bodies, organic and inorganic bodies – and the filmic products-in-production can all be called to bear witness to the peculiar ecologies of industrial society. In other words, while all workplaces might be of interest to environmental history, Ferrania is particularly so because it is an analog film factory. In its many windowless corridors and buzzing chemical labs, haptic encounter is heightened by film's peculiar affordances and sensitivities. And the film factory, which loved to

---

[13] Serenella Iovino, "Italo Calvino and the Landscapes of the Anthropocene: A Narrative Stratigraphy," in *Italy and the Environmental Humanities: Landscapes, Natures, Ecologies*, ed. Serenella Iovino, Enrico Cesaretti, and Elena Past (Charlottesville: University of Virginia Press, 2018), 69.

[14] Alaimo, *Exposed*, 102-3.

[15] Stefania Barca, "Laboring the Earth: Transnational Reflections on the Environmental History of Work," *Environmental History* 19 (2014): 3-27.

document itself on film, provides evocative visual cultural evidence of the tenor of these encounters.

One eloquent testament to the sometimes conflicting discursive and material forces at work in the factory in the late 1950s – a History of the factory that reveals the counterhistory I am unpacking here – is the documentary *Lavoro a Ferrania* [*Work in Ferrania*], an avant-garde film made in-house on Ferraniacolor film stock. *Work in Ferrania* features an energetic male voiceover, a striking electronic soundtrack, and a narrative arc celebrating the rural factory's sophisticated machinery and exquisite material mastery.[16] The film takes viewers through the process of manufacturing photosensitive film, casting this labor as one that requires "perfect control," as the script articulates. Such control, in the triumphal language of Italy's postwar economic boom, requires transcendent scientific mastery that eclipses the human. At no point do we find, on screen, collectives of humans working the line, bent over desks, gathering in the cafeteria, conducting the repetitive collective labor that underpins production. Rather, the voiceover explains as the camera pans across whirring machinery in darkened rooms, thanks to machinic innovation, "the manual intervention of man is reduced to a minimum." The film was made when the factory counted thousands of employees, and yet the narrative is one in which a disembodied human ingenuity prevails—with actual individual humans appearing only in a minor role. Throughout the documentary's twenty-seven minutes, humans most regularly appear as solitary or shadowy figures, or as disembodied details. The camera focuses on a torso or hand in close-up, for example, verifying the correct functioning of a machine, checking the quality of the product: fleshy humans are little more than machinic appendages, their limbs facilitating the flow of materials on the line.

In spite of the film's celebration of human ingenuity without human bodies, the "enmeshment of flesh with place" (in Alaimo's formulation) – and flesh with product – was fully operative in the factory.[17] A careful look at the film reveals the story of analog labor and the haptic, trans-corporeal exchange of imprints.[18] The "enmeshment of flesh with place" is most visible in the documentary when, ironically, flesh is absent. Several times, human forms appear as skeletons. The first instance is symbolic: a caution sign featuring a skull and crossbones alerts viewers (and the film crew, presumably) to the presence of radiation. Next, a pan across illuminated images of a skull and chest cavity

---

[16] The film can be viewed on the Ferrania Film Museum's YouTube channel: https://www.youtube.com/watch?v=yPmJJbRoNzo&t=1078s.

[17] Alaimo, *Exposed*, 1.

[18] Elena Past and Ed Slesak, "I rischi dell'esposizione: Pellicola e lavoro a Ferrania," in *Ecologia e lavoro: Dialoghi interdisciplinari*, ed. Carlo Baghetti, Mauro Candiloro, Jim Carter, Paolo Chirumbolo, and Maria Luisa Mura (Milan: Mimesis, 2023), 353-374.

illustrates the quality of Ferrania's radiographic film. These segments precede a section of the film that explains the meticulous daily testing of atmospheric radiation in the factory via Geiger counters and air sampling. The slightest change in radioactivity, the voiceover explains, would be devastating to the production of film. As I have argued elsewhere, these sections of *Work at Ferrania* inadvertently testify to the hundreds of above-ground nuclear tests taking place in points across the globe from 1945 until 1963, when a Limited Test Ban Treaty was signed by the U.S., the Soviet Union, and Great Britain, even though the film doesn't explain why levels of atmospheric radiation might fluctuate.[19] Although at the time *Work at Ferrania* was made, scientists were beginning to suspect that the impacts of fallout might pose health risks for humans even far from the test sites, this concern wasn't widely shared or known. The turning point for unequivocally acknowledging nuclear fallout in the context of media ecologies came decades later, with the disaster at Chernobyl.

The mutual vulnerability of film and factory workers in the face of radiation exposure makes them material kin, although in this case, film was the canary in the coal mine. Thomas Pringle's research on radiation and media recalls that "[c]elluloid and Geiger counters share a sensitivity to the material ecologies of energy that make up our world, as do human beings who are violently exposed."[20] In the factory, exposure was not "violent" but, like many industrial hazards, slow. Human bodies would record and carry radioactive traces, as well as chemical loads, over time, perhaps developing cancerous tumors or other diseases decades later. The film's delicate emulsion and light-sensitive salts, however, flagged the presence of radiation in real time. As Pringle articulates, "celluloid-based media forms [...] form a diagnostic media ecology that provides more political value than simply indexing a radiological climate."[21]

The fractures in *Work at Ferrania*'s hubristic narrative and the many points of haptic exchange come into focus through the lens of the analog Anthropocene. Alaimo insists that "the Anthropocene is no time for transcendent, definitive mappings, transparent knowledge systems, or confident epistemologies. Surely all those things got us into this predicament to begin with, where presumed mastery over an externalized 'nature' is all too triumphant, and yet also rebounds in unexpected, and usually unwanted ways."[22] Despite *Work at Ferrania's* rhetoric of mastery, film stock knew of the dangers of fallout in

---

[19] See Past and Slesak, "I rischi," 361-371. See also "Radioactive Fallout from Nuclear Weapons Testing," *United States Environmental Protection Agency*, https://www.epa.gov/radtown/radioactive-fallout-nuclear-weapons-testing, accessed March 9, 2024.

[20] Thomas Pringle, "Photographed by the Earth: War and Media in Light of Nuclear Events," *NECSUS: European Journal of Media Studies* 3, no. 2 (2014): 136.

[21] Pringle, "Photographed by the Earth," 141.

[22] Alaimo, *Exposed*, 3.

distant locales in real time. The documentary demonstrates how agentic matter inside the factory and out was always making its way in—inside the gates, into the emulsion, and into the bodies of the laborers. Film knew, and human bodies ultimately knew, the story of nuclear radioactivity that the film narrative glossed or ignored.

Sensitive Matters, or Ferrania and the *Filmiste*

I now turn to the bodies of the *filmiste,* Ferrania's women workers, in the spirit of work by scholars (like those leading the collaborative Women Film Pioneers Project) recovering the legacy of women in film production, from the silent era forward.[23] From nearly the beginning, a significant portion of Ferrania's labor force was female, and from the 1930s through the '60s, about a third of the employees—with a peak of about 1400 in 1962—were women. Many of Ferrania's departments were dominated by one gender; certain jobs, including perforation, film inspection, and packaging, were considered particularly well-suited for women, and a number worked in the factory's chemical Research & Development laboratory, primarily weighing, dosing, and sorting materials for the male scientists.[24] Recovering the contributions of the *filmiste* is important not only as a reconfiguration of film history, but also because it works against "carbon-heavy" masculinities that promote bodies invulnerable to the elements. In a way that resonates with work in feminist production studies, Alaimo argues that where environmentalism is concerned, "it remains important, since feminist histories are too often forgotten and feminist scholarship is too often marginalized, to assert the impact and relevance of feminist art, activism, and scholarship."[25]

In the Ferrania factory, the categories of *exposure* and *sensitivity* conjoin the bodies of film stock and *filmiste,* linking line worker and manufactured product in a form of material kinship—a trans-corporeal affair. Concerns about radiation in the Ferrania factory illustrate film's multiple sensitivities—which go beyond its sensitivity to light. (In Italian, rather than being called *photo*sensitive, film is often referred to more generally as "materiale sensibile"—sensitive material). Even before an image is formed on its surface, film stock archives its environment, reacting to and preserving traces of a whole host of encounters. Exposure to even a sliver of light, of course, can ruin an entire production run. But there's also the fact that dust can mar the film's surface. So can oil from human hands or dandruff from human heads. Too much sulfur in the diets of the cows from whose bones the gelatin for emulsion is made can change its chemistry, and

[23] Jane Gaines, *Women Film Pioneers Project,* https://wfpp.columbia.edu.

[24] Gabriele Mina, *Ferrania Film Museum/testi,* in collaboration with Alessandro Bechis, Andrea Biscosi, Alessandro Marenco (Cairo Montenotte: Ferrania Film Museum, 2018), 41.

[25] Alaimo, *Exposed,* 93.

thus the quality of the photographs. Bugs, molds, and bacteria can feast on that gelatin, ruining the emulsion. As we have just seen, atmospheric radioactivity can cause fogging, and trace amounts of radioactive fallout can cause spotting.

Like the film, the *filmiste* were exposed to a variety of things, and their bodies were similarly sensitive. A series of articles published in a periodical titled *Valle Bormida: Periodico dei lavoratori* [*Bormida Valley: Workers' Newspaper*] in 1961, around the same time *Lavoro a Ferrania* was made, focused on the down-skilling (*dequalificazione*) of jobs held by women in the factory.[26] Although the first several articles focus on the question of insufficient wages, the pieces increasingly examine working conditions for women in the factory. In Department D, for example, which produced reels for photography and plates for radiography, the author underlines "the intensity of the job, the physical and psychic strain, the skills required of each worker." The mechanized environment moved quickly, and in the dark, workers used their ears to monitor the correct functioning of equipment; one woman lost part of her finger in a machine.[27]

In an article about Department S.F., 15 women worked to weigh, dose, and pack chemicals to develop and fix photographs, and in the process were exposed to chemical hazards. A litany of toxic substances was present in the chemical labs: sulfuric acid, nitric acid and cyanide. To mitigate the risks of exposure, the women were required to shower at the end of their shift and use masks and rubber gloves. Yet via chemical vapors, tiny particulate matter, and other pathways, their bodies responded with varying degrees of sensitivity. The results were often painful, the article reports: spots on the skin; abrasions in the mucous membranes and mouth, which then bled; rotting skin under the fingernails; burns, sneezes, eczema, feelings of heaviness and headaches. The male workers who transported products to this department experienced the same symptoms, as well as stomach burn and gingivitis.[28]

---

[26] "Operaie dequalificate alla Film-Ferrania." *Valle Bormida: Periodico del lavoratore*, 2, no. 1 (January 1961): 2.

[27] "Vietato ammalarsi al reparto 'D' della Ferrania." *Valle Bormida: Periodico del lavoratore*, 2, no. 4 (April 1961): 2.

[28] "Escoriazioni, eczemi, gengiviti per chi lavora al reparto S.F." *Valle Bormida: Periodico del lavoratore*, 2, no. 10 (October 1961): 2. Richard Maxwell and Toby Miller's research on the Kodak plant in Rochester, New York confirms the general tenor of these accounts of toxicity, noting that by the end of the twentieth century, the Kodak plant released more pathogens into the environment than any other company; that it was releasing pulmonary and skin irritants into the groundwater; that it exposed workers to acids, acid vapors, and other irritants. Richard Maxwell and Toby Miller, *Greening the Media* (New York: Oxford University Press, 2012), 65-85.

Film's sensitivities were treated differently from the bodily sensitivities of the *filmiste*, however. In 1950s and 1960s Italy, before the modern environmental movement had taken shape, human sensitivities to bodily exposures in factory contexts were not yet fully in focus.[29] In industrial terms, human bodies were still largely treated as externalities: immediate needs were met, but the longer-term health effects of factory work were of less concern where workers were concerned. Nevertheless, the articles in *Valle Bormida: Periodico del Lavoratore* provide evidence of the fact that "one's body is interconnected with material hazards that cannot be externalized."[30] One of the factory's legacies is the lingering effect of occupational exposure across its workforce.

Routines, Endurance: Analog Film Factory as Digital Family

Beyond the physical risk to the people who labored at Ferrania, there was also a socioeconomic risk, in particular for the *filmiste*. The series of articles written for the investigative journal *Valle Bormida: Periodico dei lavoratori* were envisioned, first and foremost, to expose the low wages women earned in the Ferrania factory compared to their male counterparts—a fact that had also been reported and roundly denounced by the provincial Federation of Chemical Workers in a 1956 report.[31] Yet the factory was notoriously difficult to organize where labor was concerned, for a variety of reasons. For one, Ferrania was a paternalistic factory run, for decades, by a charismatic director whose gestures of benevolence (in particular, hiring multiple members of the same family, thus making work a truly family affair) inspired devotion and discouraged dissent. There was also the geographic factor: in the rural Bormida Valley, the dependable paychecks and regular hours gave laborers welcome relief from the uncertainties of subsistence agriculture.

The narrative I have been recounting is a familiar one about the hazards and limitations of industrial modernity. But these are not the only material, bodily, or affective legacies of the Ferrania factory. As I work to think analog Anthropocenes—the imprints of film on bodies, landscapes, and cultures—my thinking is guided by recent work by Nicole Starosielski, an ecomedia scholar who argues that although infrastructures and industries are often thought of as monolithic, in reality they exist in webs of relationships. Base infrastructures, she suggests, are in reality "fragile achievements" developed through complex,

[29] For more on Environmental Health and Safety (EHS) in the Ferrania factory, see Past and Slesak, "I rischi."

[30] Alaimo, *Exposed*, 92.

[31] Federazione lavoratori chimici Savona, "Relazione sulle condizioni dei lavoratori chimici della provincia," November 1956, CGIL Archives, Savona.

collaborative groups, and they should be understood in terms of the intimacies they require and create, not just the hazards they comprise.[32]

The parts of my research that have unfolded in conversation with former factory employees and residents of the Savona province (the broader area where the factory is located) reveal that although working hours were long and conditions not always ideal, the Ferrania factory was a respected, desirable working environment and most employees were proud of the complex, culturally significant product they helped to manufacture. Graziella Vallero (who worked at Ferrania from 1959-68) underlines that a job at Ferrania was the most sought-after career in the Bormida Valley.[33] Getting a job there, Fosca Moro (1961-74) explains, was "winning the lottery."[34]

Making analog film meant making connections and, in many cases, making lifelong friends. Vallero illustrates the intimate familial and affective bonds shaped by factory life, as one of a multigenerational family of Ferrania employees. Her grandfather managed the workshops and was known as "the German" for his blue eyes and authoritarian tendencies; her father worked in the machine room; her aunts worked in various departments, and they married other factory employees, as did Gabriella. "I was friends with everyone," she says, recalling conversations about cooking and poetry in the lunchroom. Some of those friendships, she emphasizes, are still alive 50 years later. Fosca Moro agreed that, despite the difficulty of the working conditions, it was "bello" (lovely) to work in the all-female department. At the end of the day, the women would gather in the dressing rooms to fix their hair and make-up so that they emerged from the office looking composed. Employees met lots of people—on the line, in the lunchroom, at the exits. "It was fun to be together," she recalls.

I pause over these happier details of life in the factory, because as my research has investigated questions of the slow violence and environmental impacts of the Ferrania factory, stories of joy, camaraderie, and satisfaction have also repeatedly surfaced. In my commitment to "staying with the trouble" of environmental disturbance, as Donna Haraway urges us to, I risked neglecting

---

32 Nicole Starosielski, "Environmental Media Infrastructure Practice: Bridging Production Cultures and Coastal Community Needs," roundtable talk at Society for Cinema and Media Studies Conference, Boston, MA, March 14, 2024.

33 Graziella Vallero, interview, *Ferrania a Memoria,* https://www.youtube.com/watch?v=9d-SXVX9EqY&list=PL7Dgn_AvzqnB4QTsMv_WQUfl2ieu59c6k&index=66. All citations of Vallero come from this interview.

34 Moro worked in the all-female Department G, inspecting film (as she explains) for flaws, hairs, or other problems. Fosca Moro, interview, *Ferrania a Memoria,* https://www.youtube.com/watch?v=EMyTQD80GjQ&list=PL7Dgn_AvzqnB4QTsMv_WQUfl2ieu59c6k&index=64.

the potential joy and the "necessary resurgence" in a "thick present" (or really, the recent past)—forms of solidarity that may comprise a critical means of response in an era of climate emergency.[35]

Caroline Levine's recent work on climate change, activism, and the humanities urges cultural theorists to be mindful of the places where, instead of ruptures, we find connections or continuities, and instead of aporia or absence, we find that something is being built. She suggests, for example, that humanities scholars might engage in formal reconsiderations of such things as the role of routine in human culture. Industrial labor and work on factory lines – the ultimate routine – has often, and rightly, been condemned as oppressive. And yet, Levine argues, "All cultures [...] are shaped by repetitive practices—from harvesting crops to religious observance to a weekly poker game—that keep social worlds going."[36] Routines, she recalls, do not have to "always and necessarily entail inequality."[37] Inspired by Levine to think about the affordances, and not just the perils, of routine at the Ferrania factory, I see and hear in the *filmiste*'s stories how coming together allowed for welcome forms of female sociality.

The interviews with Graziella Vallero and Fosca Moro are available in streaming on YouTube, thanks to an oral history project the documentarian Diego Scarponi conducted with students in collaboration with the Ferrania Film Museum, which opened in 2017. The archive "Ferrania a Memoria" (Ferrania from memory) contains hours of interviews with former factory employees, although just a few of these are women, and it allowed these employees to share their memories and appear in front of the camera, making them protagonists of the general genre of mediated history they participated in crafting. Ferrania a Memoria is one of many places where tender analog memories cross into the digital realm, a charged space for the factory that was shuttered when digital eclipsed analog production.

The Ferrania Film Museum additionally curates a Facebook group, "Ferrania: percorsi (foto)sensibili" [(photo)sensitive pathways], which is a virtual public square and a lively space of exchange for thousands of former employees, family members, and friends. Museum director and former Ferrania employee Alessandro Bechis moderates, frequently posts archival documents and photos, and responds knowledgably to questions and exchanges. On the Facebook page, I encountered

---

35 Donna J. Haraway, *Staying with the Trouble: Making Kin in the Chthulucene* (Durham: Duke University Press, 2016), 1.

36 Caroline Levine, *The Activist Humanist: Form and Method in the Climate Crisis* (Princeton: Princeton University Press, 2023), 58.

37 Levine, *The Activist Humanist*, 59.

the stories of Jolanda, Bruna, and Maria Rosa, which revealed another of Ferrania's analog legacies.[38]

Fig. 3.1: Jolanda "Maria," image posted to Facebook, April 13, 2022.

Permission granted by Ferrania Film Museum.

[38] "Ferrania: percorsi (foto)sensibili," https://www.facebook.com/groups/1489025068069293/permalink/2805201956451591.

Fig. 3.2: Maria Rosa, image posted to Facebook, April 13, 2022.

Permission granted by Ferrania Film Museum.

Fig. 3.3: Bruna, image posted to Facebook, April 13, 2022.

Permission granted by Ferrania Film Museum.

Jolanda, Bruna, and Maria Rosa all worked at the Ferrania factory when the *Valle Bormida* journal's investigation was unfolding, and their lives on the factory campus overlapped in time. Jolanda worked at Ferrania from October 1948 through 1969, moving between departments that included perforation and inspection of radiographic films. Bruna was hired in May of 1954 and worked at Ferrania until 1976 in departments that included cutting, spooling, labeling, and packaging film. From 1960 to 1966, Maria Rosa worked in radiographic film inspection and packaging.

In a thread on the Facebook page beginning on April 13, 2022, Jolanda's granddaughter asked Bechis if he had a photograph of her grandmother at work, since her family didn't have any. Her inquiry spurred a flurry of recollections and additional requests from other *filmiste*'s daughters and granddaughters, as well as from Jolanda's former colleagues. More photos appeared from the archives, including those of Bruna and Maria Rosa, to the delight of the virtual onlookers.

The women's striking, unsmiling faces look out from the lively Ferrania Facebook group, gazing out at their former colleagues, their daughters, curious strangers. The photo of Jolanda, taken in the 1940s, is perforated by two staples, signaling that the image was secured in her personnel file. Maria Rosa's photo is creased. Bruna's is gently spotted. These photos, scanned from personnel files recovered by museum volunteers from abandoned factory buildings, signal the beginning of the women's factory careers, as they were taken when each was hired. Their faces don't yet contain traces, whether visible or not, of the strain of working in darkness, or of contact with chemicals or pigments. Yet the photographs, stored in factory buildings for decades, stand for careers spent in contact with materials and above all with one another.

Today, the images exist in a hybrid analog-digital crossing, between factory personnel files archived in a dusty warehouse in Cairo Montenotte, Italy, digitized, and now stored on hard drives in the Ferrania Film Museum. When I saw them, they were circulating in the diffuse digital archive, part corporate, part anarchic, that is Facebook. Although their location, which gravitates between cloud servers, digital hard drives, and personal computers, is hard to pinpoint, these digital archives are not immaterial, and neither are their affects. The Facebook group is not a utopia. Between the polarizing algorithms spewed by parent company Meta, the environmental costs of server farms, and the rapid obsolescence of the devices that allow end users to access the site, Jolanda, Maria Rosa, and Bruna's visages reside in a troubled mediatic landscape. As Sean Cubitt warns in *Finite Media,* digital, "user-generated content" can never "become a universal, popular platform for innovation so long as the infrastructure that would permit it is founded on the integral waste of finite resources."[39]

Nevertheless, the exchange on Facebook transforms factory labor into a form of reproductive labor, which Stefania Barca defines as "the work of sustaining

[39] Sean Cubitt, *Finite Media* (Cambridge: MIT Press, 2018), 168.

life in its material and immaterial needs."[40] It takes the lonely personnel photo – solitary, solemn – and substitutes it with a group, restoring elements of the sociality so prized in factory life. Most significantly, the Facebook page translates the industrial knowledge these women participated in stewarding into a multi-generational dialogue. Though the daughters and granddaughters might not learn the intricacies of film inspection or perforation or photochemistry, the memories they now hold extend Ferrania's analog legacy into new affective and mediatic terrain.

Analog Anthropocene Art: Film Quilts

As the preceding pages recount, analog's legacies in the Anthropocene are complex and contradictory. In discussing her research on Kodak's industrial chemistry, Alice Lovejoy explains the complex loops of chemical elements, geographic locations, and political actors as "multi-stranded" research that requires her to weave together – without flattening – its many diverse threads.[41] This textile metaphor in the context of the photochemical industry draws me to a final example of a legacy of analog film stock in the Anthropocene, a contemporary art practice that gathers these conceptual and material threads.

The artist Sabrina Gschwandtner weaves together lengths of analog film stock to make breathtakingly beautiful film quilts.[42] Her practice began when an archivist at Anthology Film Institute called to offer her deaccessioned 16mm films from the Fashion Institute of Technology's archives, many of which were in a state of decay—suffering from fading, vinegar syndrome, or other signs of age. They were films dating from the 1950s to the 1980s, and they were all about textiles. Gschwandtner says film's repeated imagery—24 frames per second—reminds her of a textile pattern. Inspired by these materials, the images on them, and the histories of filmmaking and quilting, she began creating film quilts. She thinks of these quilts as a three-dimensional form of film editing that validates overlooked women's histories and the ways they are recorded in materials.

Her work alternately critiques or illuminates lost film histories, bringing to light underacknowledged toxic industrial practices in some cases (as in a work that features a textile factory that made camouflage for the military), or elevating

---

[40] Stefania Barca, *Forces of Reproduction: Notes for a Counter-Hegemonic Anthropocene* (Cambridge: Cambridge University Press, 2020), 6.

[41] Alice Lovejoy, "Toward a Transnational History of Extractive Media," roundtable talk at Society for Cinema and Media Studies conference, Boston, MA, March 16, 2024.

[42] "Artist Chat: Sabrina Gschwandtner," *Music for Your Inbox*, April 10, 2023, https://www.youtube.com/watch?v=lZNt79nSo4A. To see the film quilts series, visit: https://www.sabrinag.com/filmquilts.

narratives by women directors whose work has been left out of the canons of film courses. Recently, she explored the legacies of women filmmakers from the silent era, including Marion E. Wong, Alice Guy-Blaché, Germaine Dulac, and Lotte Reiniger.[43] For this project, her process involves procuring digital copies of films from archives, then having them printed at a film lab for the quilting projects. As such, it is work that materially affirms and sustains the history of the analog process, even as it relies on digital intermediaries to do so.

Although Gschwandtner's works are solo works to a certain extent, the intricate, intimate forms of the quilt patterns call attention to another kind of women's labor. In an article about Gschwandtner's work, Hugo Ljungbäck observes that quilting was "a form of speech through which the materials one worked with, the patterns one created, and the care and skill with which one stitched all inscribed values and meanings into the quilts."[44] For me, her quilts recall the collective labor my grandmother and step-grandmother convened in their rural south Texas dining rooms, when a quilt stand would overtake the dining table, and forks and knives would give way to needles and thread. Neighbors would gather around the edges of the fabric, trading gossip as they created friendship quilts, wedding quilts, patchwork quilts, sometimes destined to be gifts, occasionally envisioned as wall décor, and often simply used to keep warm.

Woven from lengths of analog film, Gschwandtner's film quilts don't keep anyone warm. They do honor labor with labor, touch with touch, aesthetic medium with aesthetic object. They stitch together a feminist history of analog labor, one that recognizes that the film histories that need preservation are not only the indexical histories imprinted on film surfaces, but also the material conditions of their creation, and the haptic and affective intimacies that enabled that creation. The film quilts recognize multi-generational knowledge at the base of modernity, treating the analog past and its makers as "partners in dialogue" rather than "exploitable resources."[45]

Gschwandtner "thinks of her own role not as to preserve these materials as they originally existed, but to 'bring into material existence what exists as possibility' in the archive.'"[46] Her work takes the peculiar affordances of film and those of quilting and creates something new: a vivid, feminist analog

---

[43] For more about Gschwandtner's work with these directors, see Hugo Ljungbäck, "Hands at Work: Patching Women's Film Histories through Sabrina Gschwandtner's Film Quilts," *Cinéma & Cie: Film and Media Studies Journal* 22, no. 39 (2022): 137.

[44] Ljungbäck, "Hands at Work," 135.

[45] Cubitt, *Finite Media*, 155.

[46] Ljungbäck, "Hands at Work," 137.

legacy that reactivates the material ecologies of women's labor. Giuliana Bruno urges that such an active view of materiality, of haptic relation, serves to refashion "our sense of space and contact with the environment [...]. Rethinking materiality in this sense, then, means fostering new forms of connection and relatedness."[47]

This chapter has proposed that the Anthropocene can be fruitfully understood by way of the reciprocal, haptic ecologies of labor in an Italian analog film factory. Film shared vulnerabilities with the laborers who made it, and sometimes it left toxic imprints in human flesh. The Ferrania factory also left behind a legacy of affective bonds and lively stories which, I propose, can be part of a strategic refashioning of that "sense of space and contact with the environment" Bruno encourages. The analog Anthropocene works against casual obsolescence, the devaluing of historical objects, processes, and figures. It recognizes that when we search out imprints and material traces, the archive offers up complex stories of lives, loves, and losses, but it also – as Gschwandtner's luminous film quilts show – provides tools for fashioning new futures.

Bibliography

Alaimo, Stacy. *Bodily Natures: Science, Environment, and the Material Self.* Bloomington: Indiana University Press, 2010. https://doi.org/10.2979/6079.0.

———. *Exposed: Environmental Politics and Pleasures in Posthuman Times.* Minneapolis: University of Minnesota Press, 2016.

Barca, Stefania. *Forces of Reproduction: Notes for a Counter-Hegemonic Anthropocene.* Cambridge: Cambridge University Press, 2020. https://doi.org/10.1017/9781108878371.

Bruno, Giuliana. *Surface: Matters of Aesthetics, Materiality, and Media.* Chicago: University of Chicago Press, 2014. https://doi.org/10.7208/chicago/9780226114835.001.0001.

Burley, Robert. *The Disappearance of Darkness.* Accessed September 8, 2025. https://robertburley.com/_rb/disappearance-of-darkness/.

Cubitt, Sean. *Finite Media: Environmental Implications of Digital Technologies.* Durham: Duke University Press, 2016. https://doi.org/10.2307/j.ctv11smp48.4.

Duncan, Pansy. "Towards a Natural History of Film Form: Silver Salts and the Aesthetics of Early Studio-Era Hollywood Cinema." *Screen* 64, no. 3 (2022): 411-426. https://doi.org/10.1093/screen/hjac041.

Environmental Protection Agency. "Radioactive Fallout from Nuclear Weapons Testing." United States Environmental Protection Agency. Accessed March 9, 2024. https://www.epa.gov/radtown/radioactive-fallout-nuclear-weapons-testing.

"Escoriazioni, eczemi, gengiviti per chi lavora al reparto S.F." *Valle Bormida: Periodico del lavoratore,* 2, no. 10 (October 1961): 2.

[47] Bruno, *Surface,* 8.

Federazione Lavoratori Chimici Savona. "Relazione sulle condizioni dei lavoratori chimici della provincia." November 1956. CGIL Archives, Savona. Ferrania a Memoria. "Fosca Moro, Interview." YouTube video. https://www.youtube.com/watch?v=EMyTQD80GjQ&list=PL7Dgn_AvzqnB4QTsMv_WQUfl2ieu59c6k&index=64.

Ferrania a Memoria. "Graziella Vallero, Interview." YouTube video. https://www.youtube.com/watch?v=9d-SXVX9EqY&list=PL7Dgn_AvzqnB4QTsMv_WQUfl2ieu59c6k&index=66.

Ferrania Film. *Film Ferrania.* Accessed September 8, 2025. https://www.filmferrania.com/.

Ferrania Film Museum. "Ferrania Film Museum YouTube Channel." Accessed September 8, 2025. https://www.youtube.com/watch?v=yPmJJbRoNzo&t=1078s.

"Ferrania: percorsi (foto)sensibili." Facebook. Accessed September 8, 2025. https://www.facebook.com/groups/1489025068069293/permalink/2805201956451591.

Gaines, Jane. *Women Film Pioneers Project.* Columbia University. Accessed September 8, 2025. https://wfpp.columbia.edu.

Gschwandtner, Sabrina. "Artist Chat: Sabrina Gschwandtner." *Music for Your Inbox.* YouTube video, April 10, 2023. https://www.youtube.com/watch?v=lZNt79nSo4A.

Haraway, Donna J. *Staying with the Trouble: Making Kin in the Chthulucene.* Durham: Duke University Press, 2016. https://doi.org/10.2307/j.ctv11cw25q.3.

Iovino, Serenella. "Italo Calvino and the Landscapes of the Anthropocene: A Narrative Stratigraphy." In *Italy and the Environmental Humanities: Landscapes, Natures, Ecologies,* edited by Serenella Iovino, Enrico Cesaretti, and Elena Past, 67-77. Charlottesville: University of Virginia Press, 2018. https://doi.org/10.2307/j.ctt2204q7x.10.

Levine, Caroline. *The Activist Humanist: Form and Method in the Climate Crisis.* Princeton: Princeton University Press, 2023. https://doi.org/10.1515/9780691250854.

Lovejoy, Alice. "Toward a Transnational History of Extractive Media." Roundtable talk at Society for Cinema and Media Studies Conference, Boston, MA, March 16, 2024.

Ljungbäck, Hugo. "Hands at Work: Patching Women's Film Histories through Sabrina Gschwandtner's Film Quilts." *Cinéma & Cie: Film and Media Studies Journal* 22, no. 39 (2022): 127-145. https://doi.org/10.54103/2036-461X/16678.

Manning, Erin. *Politics of Touch: Sense, Movement, Sovereignty.* Minneapolis: University of Minnesota Press, 2006.

Maxwell, Richard, and Toby Miller. *Greening the Media.* New York: Oxford University Press, 2012.

Mina, Gabriele. *Ferrania Film Museum/testi.* In collaboration with Alessandro Bechis, Andrea Biscosi, and Alessandro Marenco. Cairo Montenotte: Ferrania Film Museum, 2018.

"Operaie dequalificate alla Film-Ferrania." *Valle Bormida: Periodico del lavoratore,* 2, no. 1 (January 1961): 2.

Past, Elena, and Ed Slesak. "I rischi dell'esposizione: Pellicola e lavoro a Ferrania." In *Ecologia e lavoro: Dialoghi interdisciplinari,* edited by Carlo Baghetti,

Mauro Candiloro, Jim Carter, Paolo Chirumbolo, and Maria Luisa Mura, 353-374. Milan: Mimesis, 2023.

Pringle, Thomas. “Photographed by the Earth: War and Media in Light of Nuclear Events.” *NECSUS: European Journal of Media Studies* 3, no. 2 (2014): 131-154. https://doi.org/10.5117/NECSUS2014.2.PRIN.

Starosielski, Nicole. “Environmental Media Infrastructure Practice: Bridging Production Cultures and Coastal Community Needs.” Roundtable talk at Society for Cinema and Media Studies Conference, Boston, MA, March 14, 2024.

Steffen, Will, Wendy Broadgate, Lisa Deutsch, Owen Gaffney, and Cornelia Ludwig. “The Trajectory of the Anthropocene: The Great Acceleration.” *The Anthropocene Review* 2, no. 1 (2015): 81-98. https://doi.org/10.1177/2053019614564785.

“Vietato ammalarsi al reparto ‘D’ della Ferrania.” *Valle Bormida: Periodico del lavoratore*, 2, no. 4 (April 1961): 2.

Zhong, Raymond. “Geologists Make It Official: We’re Not in an ‘Anthropocene’ Epoch.” *New York Times*, March 20, 2024. https://www.nytimes.com/2024/03/20/climate/anthropocene-vote-upheld.html.

Chapter 4
On Posing

Judith Roof

Professor Emerita Rice University

Abstract

"On Posing" explores the different ways non-professional or "everyday" people who pose for photographs and models who pose professionally may both know (or not) and adjust to the differences they may perceive between analog and digital photographic technologies. Such differences in the technologies as shutter speed and length of pose, lighting, time of development, the alterability of the photographs and the photographs' likely survival contribute both consciously and unconsciously to posing practices depending on the contexts, knowledge, and purposes of the photographs.

Keywords: Digital Photography, Aesthetics, Object Oriented Ontology, Aesthetic Realism.

***

You are in front of a camera. It is a "smart" phone. You do not think twice. You gather with the others, a happy group, grinning in the direction of the phone in someone's hand aimed back towards the group. You don your group smile. There is a brief frozen moment. You can already envision the picture in your mind's eye, the happy gang looking pointedly in the same direction, as do all those who pose when the picture is a group selfie. The purpose of the picture is, it seems, always self-evident: to capture the moment.

The "frozen" moment is the pose, a still pause generally understood to produce a clear image in a camera. The processes of posing reflect the evolution of the term "pose," as it represents the etymological inmixture of linguistic sources with slightly different meanings. One meaning is from Old French "*poser*" put, place, propose 'a term in debating,' itself derived from the Late Latin verb, "*pausare*" to halt, rest, cease, pause. Both *poser* and *pausare* combine with the Latin verb "*ponere*" to put, place.[1]

The modern word "pose," thus, already represents a complex concept involving both passive and active processes: being put in place as the object of a specific

---

[1] "pose," *etymonline*, January 28, 2025, https://www.etymonline.com/search?q=pose.

viewpoint, i.e., a camera, and putting oneself in place, suggesting that both sides—photographers and their subjects—engage in defining the art of posing. The photographer with the camera and with specific ideas, assignments, and technological requirements puts the posers in place, fulfilling the needs of the photograph, the photographers' clients, and/or the needs of the posers themselves. On the other side are those who pose, who "halt, rest, cease, pause." Both sides produce and fulfill the purposes of the desired photographic product.

There are, hence, many elements that might define posing practices: the contexts of the photograph and of the pose, the posers' capacities and knowledge of the specific requirements of the photographic technology, and their expectations of the resulting photograph. Because all of these are influenced by specific photographic technologies themselves, the extent of the posers' familiarity with technological differences influences posing practices, whether that influence is conscious or unconscious. The generally less expert everyday posers may or may not be aware of most of the effects technological differences might make (other than the recent instantaneity of the digital), while professional posers—models—generally are aware and adjust their practices accordingly. What do everyday posers gain from being photographed? What aspects of posing does knowledge of the difference between analog and digital photographic technologies alter for both amateur, "everyday," posers and for professional models?

Photo Technologies

Both analog and digital camera lenses frame, focus, and capture light waves as they emanate from the scenes and objects within the scope of their lenses. Their operations, however, differ in terms of processes of development, including the time required, the complexities of the processes, possibilities for alteration, and the photographs' likely survival through time. Analog photography produces images conveyed by light through lenses that cast that light onto a light-sensitive filmic surface, producing a negative (a version of the image in which the relative dark and light tones of the image are reversed). The negative is then developed with care and precision into a positive image by means of its projection onto photographic paper in an unlit location—usually a dark room—producing the final positive photographic image.

Digital photography captures images via electronic sensors that code the captured light waves into the 256 tones that occupy the pixels (tiny points emitting light) that fill computer screens and that are reproduced on surfaces, including screens (everything from smartphones to IPads, desktops, and movie screens) and computer-printed paper.

These differences in photographic media also mean that average shutter speeds differ between the analog and digital cameras. Analog photography uses film that requires slower shutter speeds with slightly longer exposure, while digital cameras generally need less exposure time. A slower shutter speed means a longer period of stillness on the part of those who pose. Digital cameras have a broader range of shutter speeds, but tend to operate in the everyday world on rapid speeds, so those posing usually do not need to stay still more than a split second. Analog film also needs brighter light, which makes the process of posing longer and more self-conscious than posing for digital cameras. Digital photography's usual brevity, however, also reduces the necessity for prolonged formality in posing.

Another distinction between analog and digital is the time it takes for the image to be developed and available, as well as the ease with which photographs might be altered during development processes. Because of analog photography's multi-step process (with the exception of Polaroid cameras that develop film inside the camera itself immediately after exposure), developing film takes longer and is more painstaking than digital photography, whose images develop instantaneously within the camera itself. The various stages of analog film development mean that the film may be accidentally exposed to light and physical damage throughout the process. Transforming analog film is also more difficult, and thus, posers might expect fewer alterations.

Because digital photographs are developed electronically, transformations can be made easily with editing software and without any damage to the original photograph. Finally, analog and digital technologies tend to enjoy different probabilities of surviving long term, as the care necessary for analog images extends into their preservation, while the ease of the digital makes such photographs, depending on the circumstances of their production, more expendable.

What effects, then, do these differences have on the posing practices of everyday posers who are less likely to be aware of technological distinctions, and on those, such as professional models, who are familiar with the differences. Do the differences in timing, lighting, photographic quality, and alterability affect the sense that photographs record and preserve a presence in history? Despite the possible differences in posing practices introduced by technological change, there is one function that both types of photography share: that all photographs have been, at some point, records of a model's presence, and have offered a poser a moment of self-recognition. All of these photographic elements and differences come together for the posers, often as unconscious motivations for their posings.[2]

What a Photograph May Offer the Everyday Poser

Despite the extent to which someone posing for a photograph is aware of what the technical possibilities might be, all who pose will eventually look for themselves in the image, which suggests that a photograph is ultimately something akin to a delayed mirror. A photograph, even the momentary pose/snap selfie of a smartphone, is always already a fragment of history, a reminder of a previous mirror image, which offers posers back to themselves as they recognize themselves as distinct from the others who may be posing with them. In this way, posing itself, whether for an analog or digital camera, is a process of staging an event in which posers can recognize themselves as the objects/subjects of inscription in a way that enables a certification of their past experience as selves. Everyday posers may also perceive photographs as a species of objective proof of co-presence in a specific environment, capturing a moment and demonstrating their presence.

This self-recognition is a process experienced by anyone who poses. Although the resultant photograph is not literally a mirror either of its subject or of the photographer, it does operate in a way similar to the "mirror stage" moment defined by Jacques Lacan as a key moment in subjective development.[3] Not only is the setting of any photograph a species of "stage" (in a bad play on words), the photograph also hosts a moment in which those who view themselves in photographs again recognize themselves as separable beings. This rediscovery of self as subject / object iterates the moment in subjective development, when, according to Lacan, the "mirror stage" occurs, which operates "as a particular case of the function of the *imago*, which is to establish a relation between the organism and its reality—or, as they say, between the *Innenwelt* (inner world) and the *Umwelt*" (world as perceived by the organism)."[4] Subjects are subsequently able to identify themselves as selves because they have already

---

[2] "Film vs. Digital—A Photo Comparison," *The Darkroom.com*, January 30, 2025, https://thedarkroom.com/film-vs-digital-comparison/?srsltid=AfmBOorjGOBQpcoeCLN3tbL1UTbSxh884v9r-n9NJzcO2l9XoiguK1qS; Roland Barthes, *Camera Lucida: Reflections on Photography*, trans. Richard Howard (New York: Hill and Wang, 1981).

[3] Jacques Lacan comments: "We have only to understand the mirror stage *as an identification*, in the full sense that analysis gives it: namely, the transformation that takes place in the subject when he assumes an image" (2). Jacques Lacan, "The Mirror Stage as Formative of the Function of the I as Revealed in Psychoanalytic Experience," in *Ecrits*, trans. Alan Sheridan (New York: W.W. Norton, 1977), 1-7.

[4] Ibid, 4.

experienced the mirror stage's initial moment of self-recognition as a separable subject in relation to an exterior world.

The process of posing for and then seeing photographs of oneself is both enabled by the mirror stage and iterates that stage in a confirming (or sometimes perhaps a not-so-confirming fashion). Re-recognizing oneself as the "subject" in a photographic "object" constitutes a process through which subjects simultaneously apprehend and objectify themselves, a process enabled by the relative permanence of the photograph itself. Photographs also reconfirm subjects' recurring self-recognitions—recognitions, which ironically, are not mirror images at all, but precisely images of the selves they can never see except through inscriptive processes that involve the apprehensions and techniques of a go-between. As Lacan notes, "the mirror-stage is not simply a moment in development. It also has an exemplary function, because it reveals some of the subject's relations to his image, insofar as it is the Urbild of the ego."[5]

As with the mirror, photographs provide the "other" surface/process necessary to subjective self-apprehension, but in a way that preserves the moment, both of the subject's existential function as poser and as the subject who apprehends itself as both a separable subject and as a subject who sees itself being seen as an object of vision. These processes occur both consciously and unconsciously for posers, who, in seeing any ensuing photographs, simultaneously recognize and identify themselves as subjects and as visible, separable objects and judge their appearance, often either experiencing satisfaction or disappointment (or sometimes both).

Offering a species of "objective" truth of their co-presence in a specific environment, the photograph, though it has immediately lost the moment itself, seems to have captured time. This certification of co-presence in time constitutes one of the primary motivations for being photographed—the desire to capture and hold onto moments, proof of co-presence in time and place.

The Everyday Poser

The average person, whose knowledge of photographic technologies might extend both to the differences in timing and permanence between photographs taken with a dedicated camera (either analog or digital) and a cell phone and the difference between having to have photos developed and having them appear instantly on the photographic apparatus itself, is less likely to be aware of the more subtle technological differences between analogue and digital apparatus. For such posers, the contexts and circumstances of the photographic moment are more likely to define the pose.

---

[5] Ibid, 4.

Insofar as digital technology, with its immediacy, has supplanted the analog in most social situations, the appearance of analog photographic equipment (camera, tripod, lighting) motivates a different posing scenario for everyday posers. Photographs taken with analog cameras have long offered the sense that photographs offer a lasting documentation of a past moment as well as of history. Posing for photographs taken with analog cameras is a conscious and often serious process that requires at least a momentary, self-conscious pause on the part of the poser, in so far as an analogue camera requires a longer still pose on the part of participants, a certain, consciously artful arrangement of people and objects, and often (though not always) a circumstance that deserves photographic record. Those who pose for the kinds of everyday digital photographs people take for fun would currently only be aware of this analog difference if a camera (whether analog or digital) sat atop a tripod or the photographer instructed participants on how they should pose.

With an analog set-up, many posers might become more self-conscious, more aware of the onset of a more involved process with a camera whose sole purpose is to produce photographs, something that in the current technological climate only a serious photographer would use. Older everyday posers would have more experience with analog photography, and perhaps wonder whether the camera has film, whether the film will need to be developed, what the photographs' quality will be, and how faithful copies of the photograph might be, which also may inflect the way they pose.

Digital Inflections on the Everyday

The much more informal poses often performed for digital selfies are enabled by the rapidity and relatively low cost of digital photographs, even if amateur posers really have little idea of the digital technology beyond that of the computerized character of the now universal presence of the cell phone. Selfies, for example, are contextual, and those who pose have some notion of a selfie preserving the social moment it captures. In the groups posing for selfies, however, there may be a poser who does not want to pose or is afraid of what the image will show. Those posers pose as non-posers. While some selfie subjects overpose, competing for space in the image, brilliance of façade, and other modes of brightness which draw immediate attention to the subject in the resultant photograph, other posers seem to blench at the edges, often merely looking at the camera, appearing as if they do not want to be seen. Even as the postures of these posers may indicate shyness, insecurities, or an unwillingness to enact the species of performance posing (even for a group selfie) requires, non-posing is still posing. It is posing as if posing means nothing to the poser, which might in fact indicate more ego than insecurity. In any case, the posers 'modes of posing enact their relation to their own images, which in turn project the complex relations the posers have with those images as the prototypes of their egos—of their sense of themselves.

Some reluctance to pose, however, may also derive from the current omnipresence of covert surveillance technology, which began with security video cameras and currently has become much more omnipresent, as digital movie cameras now oversee many public and even private spaces in the name of security. Most contemporary citizens know these cameras are present and operating, but they usually forget the cameras are there. When they are conscious of being recorded, people do begin to pose. To some extent, because of the omnipresence of these surveillance cameras, people have become full-time posers in public spaces. Those who object to surveillance also pose, evincing more covert and less attention-drawing behaviors (which, ironically, may draw attention). This surveillance paranoia (which probably is not a paranoia, since in fact, the surveillance is generally present) affects modes of posing in more open circumstances, as, for example, when one walks behind a group posing for a selfie, or perhaps a photographer taking pictures of crowds. Those who are aware will generally pose in a manner consistent with their own relationship to their egos, cowering, smiling, drawing attention to themselves, or trying to become invisible.

Retouching History in the Everyday

Some who pose for photographs also know (even if they are not conscious of the knowledge) that photos can be altered—retouched—often to improve the final photographic product for a market, as in advertising or fashion photography. Professional models are aware of these processes—and in fact, count on them to correct any flaws in appearance, posture, facial expression, or lighting, among other elements. Although photographers are also able to touch up analog photographs, digital photography has made the process of post-shoot emendation much easier, offering greater facility and breadth of possibility for alteration.

Hence, the sense that amateur posers may have that photographic processes preserve some moment in time comes from the erroneous survival of the more likely unaltered permanence of analog photographic processes. Digital photography's simulation of analog photographs (at least to the naked eye) unconsciously affects the posers' sense that posing is a way of preserving not only a second of the poser's co-presence through time, but also some real aspect of the poser's existence. This feeling of the poser's continued presence has made photographs cherished relics, pictorial evidence of the existence of the photograph's subject.

Contemporary posers, however, might also have an inkling of another aspect of digital technology: that the photograph will in fact not survive beyond the admittedly short lives of the technology (software, specific machines, etc.) unless the image has been printed. And even if that is the case, the printed media deployed by digital technologies are much less likely to survive in any way commensurate with earlier analog photographs. Because digital photos are developed by programs, they do not always represent scenes accurately, altering colors to fit into their 256 tones and misrepresenting depth relations. Digital photographs are also much more likely either to be stored in a machine's memory, which will fail with the inevitable failure of the machine, or will no longer translate into newer coding systems. Nor are digital photos likely to be printed on high-quality photographic paper that might extend their survival. There may be, thus, an inkling that photographs no longer reliably provide any long-lasting evidence of existence; and hence, the poses in the end may bear less weight, are finally nothing but minor evidence of anything other than a passing moment. In this case, the pose itself becomes less important. However, for the tenure of their persistence, digital photos may be easily reproduced and distributed.

Model Posers

For professional models, posing is performance, cameras are the method, and photographs are the media. Since models do perform in front of cameras, they are acutely conscious both of the capacities of the photographic technologies with which they are working and of the appearances and methods of posing in relation to those technologies. Professional models adjust their performance in the contexts of productions: studio, live action (which simulates images akin to the opportunistic photos of cell phones), both of which will likely be edited, perhaps altered, and adjusted to serve the commercial purposes of the shoot.

The goals of commercial photography include not only the production of stylish photographs that incite consumer desire, but also the production of images that stimulate and fulfill imaginary desires and identifications. As Heidi Klum comments, "As soon as the music starts thumping and the shutter starts clicking, we know—if we're good at what we do—how to sell it. What does 'selling it 'mean? I think it means playing at the person I want people to believe I am" (i.e., playing at the person I want to see when I look at my own picture).[6] Further, Klum describes models' performances as the ability both to produce and to project the feeling the photograph requires precisely as an appearance accompanied by the cues of ambiance and desire. As she comments, "[t]his ability to project an image is a trick of the trade. If a photographer poses me on

[6] Heidi Klum, *Heidi Klum's Body of Knowledge: 8 Rules of Model Behavior* (New York: Crown Publishers, 2004), 42.

a bed and tells me to look sexy, I turn into an enchantress, beckoning to someone just off camera. Or more accurately, I become an actress playing that siren."[7]

On the other hand, a model can try too hard, overplaying the assigned attitude and overposing. While everyday posers anticipate seeing themselves, professional models anticipate seeing themselves performing. For professional models posing shifts from simply pausing to being placed, or "put" in a specific pose. What professional models see in commercial photographs are images of themselves enacting roles, postures, styles, and social positions that represent specific presentations of gender, age, social class, sexual attraction, elegance, self-confidence, and other strategies contrived to attract customers to the products they display as a part of their profession. They are aware of the specific requirements and limitations of photographic technologies in the staging, production, and reproduction of images and pose accordingly, usually as directed. Although their task is to achieve various appearances, as Klum suggests to potential models: "Your style is an extension of your personality."[8]

The posing methodologies of models are also designed to offer what appear to be images of the factual presence of style and utility in relation to clothing, automobiles, and other consumer items. The motivations of the styles of posing point to photography's ability to deceive while delivering authentic images of presence, even as photographs may be retouched, altered, recombined, or are the product of any one of a number of potential processes of alteration. As Klum comments, "It's funny how you can become practiced at creating illusion thanks to the magic of photography, the three staffers were cropped out of the picture, and I looked like I was just luxuriating in some balmy water."[9] Model Tyra Banks notes that "My years in the business have taught me that the camera doesn't lie. If I have a pimple, a dark spot, or a rough patch, it will definitely show up in a photograph. And while it's true that photographers do use retouching to erase noticeable blemishes, too much retouching can become expensive."[10] Although analog photography does enable alterations in photographs, digital technologies offer far more extensive and much cheaper means for altering what appears to have been before the camera lens.

Finally, as Klum envisions the processes of posing, modeling is ultimately a form of "communication" with the camera and is finally "about desire," a concept that is related not only to consumers, but also to the models who

---

[7] Ibid, 42-43.

[8] Ibid, 71.

[9] Ibid, 32.

[10] Tyra Banks, *Tyra's Beauty Inside and Out,* (New York: HarperCollins, 1998), 20.

experience the effects of Lacan's understanding of the subject's relation to its own image in so far as seeing one's image is about a desire to find (or refind) self.[11] The deliberate appearances professional models project are successful only insofar as the model can merge performance with sincerity, at least as models who comment upon the modeling processes suggest. As Klum notes: " . . . physical beauty, and the allure that someone projects through facial expression and demeanor, are all things that can be worked on, improved, embellished. Of course, there's no substitute for that real quality that draws people to you—charisma, sex appeal, warmth, exoticism, confidence, shyness—but sometimes it takes practice just to let go and allow these qualities to shine through . . ."[12] Klum envisions the processes of being photographed performing as a model ultimately as an aspect of the model's quintessential self (as in the "mirror stage"), since in the end performers both project and discover the truth of their being. It is also, as Klum suggests, a matter of measure—of not going too far and of knowing what to display and what to hold back: "The secret to beauty is to highlight and hide. Highlight not only what you want them to see, but to the point that you've distracted them from what you *don't* want them to see."[13] She avers, "Your style is an extension of your personality."[14] Klum concludes that "professionally, it's my job (among other things) to simulate simulation, fanning the fantasy that anything is possible: that I can be yours, that you can be me—heck, that I can be me."[15]

That posing professionally is both "simulation of simulation" and the presentation of a "real" self suggests that the processes of posing are both ambivalent and self-contradictory. A model must simultaneously perform the appearance and feelings that produce the sought-after effects, while manifesting aspects of the more genuine "self," managing the appearance, appearing as what photographers want her to be, while both discovering and remaining true to herself. This suggests that photography, whether analog or digital, is essentially two-faced, both insofar as it presents a simulation and simultaneously offers some insight into the character of the poser.

In this context, analog photography's recording of reflected light rays from actual physical bodies in front of the lens would seem to capture all of what is before the lens and frame: appearance, simulation, and evidence of some core persona. Analog images might be touched up, but with analog photography,

[11] Klum, *Heidi Klum's Body of* Knowledge, 14.

[12] Ibid, 60.

[13] Ibid, 62.

[14] Ibid, 71.

[15] Ibid, 138.

touching up consists of blocking, removing, or altering problem areas with a brush and retouching fluid and/or tone dye. The process is manual, performed by one who physically develops the photos. What the process of retouching suggests is that the development processes of analog photography afford another potentially creative step that ultimately can affect the poser's appearance—a step that is completely out of the poser's control. Touching up digital photographs is a process equally, if not more, beyond the poser's control insofar as digital touching up occurs using software fixes with the capacity to add and subtract far more radically than is possible with an analog photograph.

Hence, the photograph's fact/fiction ambivalence also becomes a question of the extent to which any photograph has remained a consistent "record" of the original scene. The quality of "simulation" comes not only from the poser's concurrent manifestation of performance and self, but also from potential alterations made during the development, review, and final appearance of the photograph, whether that is in print, on-line or both. When they might review their own post-production images, professional posers are much less likely to look for themselves as selves as a reaffirmation of their subjectivity and more likely to appreciate their sincere capabilities to provide perfect simulations—themselves removed from self-as-self to self-as-performers. Appearance as opposed to substantive existence is what counts, which may be why Klum insists upon there being some manifestation of authentic posers' personal qualities shining through and sustaining the simulation as a multi-layered image, whether analog or digital.

The notion of simulation also suggests a layering in the image, not only points to retouching processes, but also to analog technologies as a sustained originary model for photographic production. Analog photography depends upon the layers of chemical light-sensitive material whose alterations upon exposure preserve the moment in which the film was exposed to light through the camera lens. The sense of capturing a moment persists in digital photographs, even though the moment is immediately transformed into a mass of circuits whose off/on coding merely represents a moment whose physical traces are no longer present. The digital ultimately reduces the complexity of light and color captured to what the coding itself can produce, a surface expressed through the number of pixels available on the electronic screen. Although the differences are barely visible to the human eye, the instantaneity and alterability of digital encodings affect posers 'sense of permanence as well as their sense of what photography ultimately captures. In a digital context, photographs increasingly become raw material for a final product that may or may not look like the moment of the original capture. And in the end, professional models are more likely to be aware that posing is, finally, not the end of the posing process insofar as digital technology can recontextualize, resize, alter color, and change many aspects of the context and environment of the original pose.

Poseuses: Photographic Portraiture

Because analog photographic processes tend to be slower, those who pose for analog portraits generally pose thoughtfully, deliberately, and self-consciously, and in art photography, the posers become objects defined wholly by the photographer. In early analog portrait photography, posing often took long periods of time, twenty seconds by the 1840s.[16] Even as analog film became more sensitive and posing times became shorter, portrait photography still required self-conscious arrangements of clothing, posture, facial expression, and lighting. A good example of this is Gon Buurman's *Poseuses: Portraits of Women* (1987), a collection of analog portrait photographs representing a variety of Dutch lesbian women from the 1980s.[17] All of the photographic subjects are consciously posing, but the printed book is a collection of posers who are not professional posers posing for a professional photographer, Buurman, not simply for their own portraits, but for a collection of portraits that represents a specific group of individuals.

In *Poseuses*, the photographic subjects are both posing and not posing insofar as they are simultaneously striving to present the most favorable portrait of themselves as well as the most quintessential example of lesbian style and essence. As Pamela Pattynama suggests in her introduction to the volume, "Photographs sometimes reveal what is otherwise so carefully concealed."[18] Since part of the purpose of the collection is a documentary presentation of the character of actual lesbian denizens, the desire, whether conscious or unconscious, of those who pose is to present themselves as they actually are, which in itself involves both some sense of fact and of fantasy, of conscious and unconscious postures. As Pattynama comments: "Many-sided, many-minded and paradoxical, the lesbian sensibility gives itself away in photographs. Unrestrained by principles, it crops up at unpredictable moments, only to recede again into the next metamorphosis" (46). Sometimes the poseuses are overtly posing, as do "[t]wo women who parody feminine glamour with a pastiche of the thirties. They are clear poses. The pictorial effect does not admit any deeper truth."[19] However,

---

[16] "In Focus: The Evolution of the Personal Camera," January 31, 2025, https://dp.la/exhibitions/evolution-personal-camera/early-photography.

[17] Gon Buurman and Pamela Pattynama, *Poseuses: Vrouwenportretten/Portraits of Women*, trans. Sheila Gogol, (Amsterdam: Uitgeverij An Dekker, 1987).

[18] Buurman and Pattynama, *Poseuses*, 46.

[19] Ibid, 29.

Pattynama continues, " . . . the show, the irony with which a woman presents herself often suffices."[20]

The Posture of Digital Facility

Analog photography's ability to capture depth beyond surface, as well as the relative permanence of its images, sustains photography as an art that consists of the presentation of visions, often imaginary scenarios made real by having been photographed. As an artistic medium, analog photography produces images that no longer represent attempts to capture aspects of a poser's appearance or to document events. Instead, the technologies associated with photographic lighting, models who pose in sculptural presentations, framing, exposure lengths, and processing provide multiple means for transforming staged scenes into more transcendent images. Sadly, these analog attributes will slowly disappear.

The coded nature of digital photographs makes such photographs and portions thereof much easier to manipulate: to move, resize, and/or insert into a completely different context in another coded file. Just as digital photographs are much more easily distributed yet much less permanent, so they are much more easily alterable. At the same time, despite the difference in the photographic technologies, when the context of posing appears to be the same as portrait photography with what appears to be an analog camera, amateur posers will continue to pose the way they have always posed for the camera. In this context, digital cameras and formal portrait contexts imitate the posing techniques of the analog, perhaps motivating little, if any, difference to those who pose.

The digital product and processes, however, can produce scenes that look like photographs of scenes that never existed. The reliability of photography as that which captures factual events disappears with the digital. Thus, professional models who pose in the era of digital technology may well do so only to provide raw material for something beyond portraiture. Photography has shifted from a co-present witness to an element in a coded, artificial and fictional scenario. For those who pose and who are aware of this shift, this is the difference between the preservation of self as presence and the temporary manipulations of coded lights whose alterability and transferability take photography from witness to manipulable and often temporary simulations of simulation with a life span that lasts as long as the current version of the technology.

Bibliography

Banks, Tyra. *Tyra's Beauty Inside and Out.* New York: HarperCollins, 1998.

---

[20] Ibid, 29.

Barthes, Roland. *Camera Lucida: Reflections on Photography.* Translated by Richard Howard. New York: Hill and Wang, 198.

Buurman, Gon and Pamela Pattynama. *Poseuses: Vrouwenportretten/Portraits of Women.* Translated by Sheila Gogol. Amsterdam: Uitgeverij An Dekker, 1987.

"Film vs. Digital—A Photo Comparison." The Darkroom.com. January 30, 2025. https://thedarkroom.com/film-vs-digitalcomparison/?srsltid=AfmBOorjGOBQpcoeCLN3tbL1UTbSxh884v9r-n9NJzcO2l9XoiguK1qS.

"In Focus: The Evolution of the Personal Camera." January 31, 2025. https://dp.la/. exhibitions/ evolution-personal-camera/early-photography.

Klum, Heidi. *Heidi Klum's Body of Knowledge: 8 Rules of Model Behavior.* New York: Crown Publishers, 2004.

Lacan, Jacques. "The Mirror Stage as Formative of the Function of the I as Revealed in Psychoanalytic Experience." In *Ecrits A Selection,* 1-7. Translated by Alan Sheridan. New York: W.W. Norton, 1977. https://doi.org/10.4324/9781003209140-1.

"pose," etymonline, January 28, 2025, https://www.etymonline.com/search?q=pose.

Chapter 5
Compelling Abstraction: Cynthia Greig's Digital Transformations of Art Space

renée c. hoogland
Wayne State University

Abstract

Organized around several works by Michigan artist Cynthia Greig, the chapter provides an exploration of contemporary digital art photography in relation to the aesthetic act through an engagement with object-oriented ontology and the new aesthetic realism.

Keywords: Digital Photography, Art Spaces, Object Oriented Ontology, New Aesthetic Realism.

***

> In the end, writing and photography probably destroy more of the past than they ever preserve of it.
>
> Sigrid Nuñez, *The Friend* (2010).

I am looking at a winter landscape, opening up to the horizon. While bright, the light has disappeared from the sky, hidden behind a solid mass of clouds, on one of those days just after Christmas, when the world seems to be standing still for a moment, expanding into the massive space of sky, brilliant but opaque. In the foreground, a surface stretches to meet the deceptive demarcation line between earth and space that, according to family lore, provoked me at age four, on a day at the beach, to observe to my mother: "In the distance, the sky is lying on the ground." The image not only reminds me of but also reactivates this early perception. The stretch of land-ground, brownish-grey, with several greenish patches cutting across the slightly curved ridges that mark its surface, positions me, the beholder, in a landscape that feels familiar, that enfolds me, that I can almost smell, even though the surf line is too far away to make me hear more than a slight buzz behind which I assume the crashing waves to be. The label attached to the wall under the work, which is clearly a photograph but which, because of the textured quality of its subject, also evokes the tactility of painting, reads: Matthew Marks/*Brice Marden: New Paintings*—and therewith begins to confuse my perceptual experience.

Fig. 5.1: Matthew Marks, Brice Marden: *New Paintings*, (Chelsea) archival pigment print mounted on Dibond, 14.5 x 22 inches, 2012/2014.

The image described is part of the series *Gallery Horizons*, a set of digital 14.5"X22" photographs by Michigan artist Cynthia Greig. Walking through the exhibition, looking at the other works in the series, and pondering the labels, all including the name of a gallery, an artist and the title of their show, it gradually dawns on me that I am not actually looking at landscapes. The exhibition notes explain that *Gallery Horizons* consists of close-ups of the space where the floor of a particular gallery meets the wall. Each work constitutes a material visualization of typically invisible spaces that are digitally transformed, abstracted to horizon lines, creating space and thus actualizing deep landscapes that emerge only in the perceptual encounter, in the event of looking—looking, as it turned out, elsewhere, into imaginary depths.

Fig. 5.2: Skarstedt/Justin Adian/New York, archival pigment print, 26 x 40 inches, 2015/2018.

A second, related series of Greig's recent work, equally, albeit in a very different way, plays havoc with and therewith confuses conventional practices of looking inside the traditional "white cube" of the contemporary art gallery. Titled *Threshold*, these large-scale (26.5"X40") color prints pull me into long-shot views of gallery spaces in which the walls are startlingly empty: the artworks have been digitally erased. Foregrounding the typical architectural style of the contemporary art gallery—high ceilings, straight angles, pristine white walls, and shiny grey concrete floors in which traces of the erased artworks occasionally appear reflected—the photographic image contains no artwork but people, alone or in smaller or larger groups, seen from the back, "dwarfed" by the cavernous gallery space, looking up at empty white walls, ostensibly unaware of the fact that they are being transformed from subjects into the objects of my look, from lookers into that which is being-looked at. These photographs do not merely throw the subject-object relation of the gaze into disarray; however, they also deliberately problematize the act of looking in relation to art in its dedicated spaces and therewith raise the question of aesthetic perception per se—a question, I submit, that is central to the complexities of the photographic image in the wake of the digital turn.

Fig. 5.3: Pace/Nigel Cooke/New York, archival pigment print, 26 x 40 inches, 2015/2018.

Gallery Horizons and *Threshold* are both unabashedly digital projects. Greig's determined attempts to "dislocat[e] the viewer," to throw any straightforward, historically naturalized "single point of view" into disarray by "expos[ing] the unexpected forces of entropy that operate beneath the art intended to be our focus," forced me to, and I quote Greig again, "meditate on the idea of Nothingness" as at once expressed in and denied by these photographs, shot in art galleries and museums across the US and Europe.[1] While not exactly engaging me in what I have elsewhere called the violent embrace of the involuntary forces of certain cultural practices—Greig's work, I posit, is at once entrancing and cerebral, not the most "theatrical" kind of art so forcefully dismissed by art historian Michael Fried; they are, what one might call, "low affect"—these photographs nonetheless pull me into an embrace that involves at once feeling and thought, both sensation and reflection. In this way, they not only stage an aesthetic encounter but also force contemplation about the nature of such encounters within the architectural conventions of the white cube of the contemporary art gallery. By fundamentally reorganizing the relationship between the subject and object of artistic visuality, the photographs therefore alter the field of aesthetic perception as such. In the context of this collection of essays, they invite me to explore ways in which the digital image actualizes visibility where none was before, in particular in the realm of

[1] Cynthia Greig, *Gallery Horizons* (artist statement), 2018. n.p. https://www.cynthiagreig.com/cynthiagreig/art/galleryhorizons/ghindex.html.

aesthetic production and its reception, in the domain of art in its "worlding" functions.

At first glance, the photographs in *Threshold* appear to literalize what the founder of Object-Oriented Ontology (or triple o), Graham Harman, suggests to be the essential and conditional quality of artworks generally. Assuming aesthetics to mean the "general theory of how objects differ from their own qualities," Harman defines beauty as the "disappearance of a real object behind its sensual qualities."[2] Rejecting Kant's central "distinction between *phenomena* and *noumena*, also known as appearances and the thing-in-itself,"[3] as well as any phenomenological approach ensuing from it, OOO emphatically reaffirms the thing-in-itself as not only the "key to future progress in philosophy"[4] but also the basis on which aesthetics establishes itself as "first philosophy."[5] In his recent book on art and objects, Harman develops this idea by, first, claiming that an artwork is defined by the fact that it "cannot *primarily* be a form of knowledge" and, second, while potentially generating a "cognitive activity" as a kind of "side effect," is therefore in the final instance "unparaphraseable."[6]

Harman posits that knowledge statements are literal statements and "literal statements treat objects ... as interchangeable with a list of qualities it possesses."[7] Artworks, sensual objects with real qualities, contain non-literal elements, which is what qualifies them as art: describing their physical composition does not tell us much about the operations of such objects qua art, but neither do artworks rely on some vague form of aestheticism. In order for aesthetic effects to be created by a work of art, and hence, for beauty to emerge, the real object, Harman writes, must withdraw, disappear from the literal level, without nonetheless becoming inaccessible. On the contrary, the disappearance of the real object behind its sensual qualities is precisely that which Fried dismisses as a theatrical effect, but that which Harman identifies as the conditional quality of art. This effect, Harman argues, cannot be approached directly, cannot be, in Gilles Deleuze's alternative terms, be brought under the reign of the signifier, but it can be "allude[d]" to;[8] the artwork, then, not being

---

[2] Graham Harman, *Art + Objects* (Cambridge UK: Polity Press, 2020) 24. Further references to this work will appear in parentheses in the text.

[3] Harman, 13.

[4] Harman, 14.

[5] Harman, 31.

[6] Harman, 30.

[7] Harman, 26.

[8] Harman, 30.

a form of knowledge yet real, both as a thing-in-itself and in terms of its sensual qualities "need[s] to be approached from the side rather than head-on," he concludes.[9] It is the non-literalness of the work of art that defines it as such. The work of art, in order to do its work, then, would appear to be both dependent on the viewer or beholder, as much as it relies on its own theatrical properties to be engaged with, to come into its own qua art, from the side.

Fig. 5.4: Paula Cooper/David Novros/New York, archival pigment print, 26 x 40 inches, 2017/2018.

In literally erasing the artworks from the gallery walls in *Threshold*, Greig gives concrete, material form to what Harman calls the withdrawn object that gives rise to the theatrical effect of aesthetic experience. What we see when we look at these photographs is not the work of art that has been erased from the wall, but the beholder's engagement with what has been turned into an empty space. In other words, it is precisely in their absence that the artworks elicit our attention to what is not merely no longer there, but to what is essential to the aesthetic experience per se. The photographs furthermore explicitly identify artworks, paradoxically removed from the gallery walls, as the site where the inscrutable real object hiding behind its sensual qualities emerges and is supposed to become accessible, available to the beholder, albeit obliquely. At the same time, the photographs set in motion a reversal of the traditional visual *mise-en-abyme*. The empty white space of the walls—a space of literal absence—where we expect the artworks in the gallery to appear, is turned into a space of

[9] Harman, 31.

presence by and for the act of looking—not only the looks of the faceless figures observing the walls, expectantly looking up, but also our own looks as beholders of the photographs. On the literal level, the gaping white walls—digitally enhanced to retain the texture and detail of the actual gallery walls—appear to lure us into the position of the viewer beholding those who are looking at materialized absence. Our gaze seems to be redirected from the invisible objects of the look (the artworks) to the subjects of this type of "institutionalized" looking, yet we do not actually see these looks nor the subjects' eyes. Instead of, as I have suggested before, turning the already transformed subjects in the photographs—the people in the gallery spaces—into the objects of our gaze, each photograph in the series—and the effect of the repetition is quite relentless—is precisely, in the terms of OOO, to draw us to the withdrawn object, the inscrutable object of the photograph, the object-in-itself that it at once creates and obscures, and that, Harman insists, forms the act of aesthetic experience as such.

Still, since the object cannot be separated from its qualities—they always "come as a pair," Harman aptly observes[10]—something needs to be left or happening in its wake. In a quite remarkable move in view of his firm dismissal of Deleuzian notions such as process, becoming, the virtual, and the event in his earlier work, Harman, in *Art + Objects* (2020), puts the beholder of the artwork in this place. Yet, whereas Kantian aesthetics situates beauty exclusively in the eye of the beholder, OOO rejects the opposition between subject and object of the aesthetic act as such and insists on the beholder as the "real object that performs and sustains" the artwork's qualities once they are "stripped" from the actual object,[11] while nonetheless insisting that "each of the terms has an autonomous reality not exhausted by its relation to the other." Simply put, the "artwork exists only as a hybrid of work and beholder,"[12] and it is as a new object that this hybrid persists. Harman here appears to come very close to what Deleuze calls the aesthetic encounter, but for the fact that the object-oriented philosopher draws a clear distinction between the human being as an "*ingredient* of art and as a *privileged beholder* of it." The new object, which like Deleuze, he defines as a "compound," is more than the sum of its parts: it "exceeds both parts individually and is not exhaustively knowable by the human beholder who forms part of it,"[13] and therefore cannot be defined in terms of what it does, but must be understood as an object, as what it is.

---

[10] Harman, 29.

[11] Harman, 29.

[12] Harman, 69.

[13] Harman, 45.

While I consider OOO's theory of aesthetics helpful and overall persuasive, I find myself struggling with the latter idea on several grounds. First, Harman's denial of any kind of doing to the aesthetic act appears to run counter to his own invocation of the beholder as someone or something that needs to "stand in for the real object that disappears from our grasp."[14] Indeed, it is precisely the active participation of the beholder in the work of art that leads him to the conclusion that there is a "primordial theatricality to all the arts,"[15] and, as we have seen, that it is the reader or beholder who "performs" and "sustains" the object's qualities in its absence. Clearly, even if the "unified reality" of the work-beholder hybrid cannot be said to be "exhausted by any particular understanding of it," this does not in any way mean that the work itself is not also doing something, i.e., in calling forth the beholder, upon whose active, theatrical participation it depends. This would most definitely mean that there is an aspect of doing involved in the viewer's engagement with the work of art, whether as "ingredient" or as "privileged beholder" of it. Second, Harman's theory of art in terms of withdrawn or inscrutable objects that disappear from our perception in our attempt to grasp them, paradoxically, would appear to entail that the being of the object as such is not so much obscured as it turns into a logical impossibility. Indeed, as German philosopher Markus Gabriel argues in his short book, *The Power of Art* (2018),[16] in which he defends the radical autonomy of art within what he terms, a *new aesthetic realism*, OOO's notion of the withdrawn object means that "aesthetics can never put us in touch with art as it really is or as *art in itself*." In other words, Gabriel continues, the object's disappearing act renders art itself into something that is "not really part of reality but only a scene celebrating our alleged incapacity to perceive things as they are."[17] In the space that remains me, I want to foreground Gabriel's New Realist model of art, in particular his notion of "fields of sense," in order to gesture towards a possible approach to the enigmatic and compelling abstractions of Greig's *Gallery Horizons.*

Gabriel starts his defense of the power of art from the straightforward assumption that "[w]e can perceive objects because we exist in the same domain" or, in what he terms the same "field of sense."[18] Fields of sense are an "ensemble of objects which [sic] appear in a certain manner"[19] (24) or,

---

[14] Harman, 71.

[15] Harman, 71.

[16] Markus Gabriel, *The Power of Art* (Cambridge UK: Polity Press, 2020). Hereafter *PoA.*

[17] Gabriel, *Power of Art*, 28.

[18] Gabriel, *Power of Art*, 24.

[19] Gabriel, *Power of Art*, 24.

alternatively, as he puts it elsewhere, "the places in which everything appears."[20] Fields of sense are not "object domains" because "what makes a field of sense is not exhausted by its being a field of sense."[21] Gabriel contrasts a field of sense with a room, which remains a room whether there are objects in it or not. Fields of sense, however, are "not to be understood without the orientation and arrangement of the objects which appear in them" and are therefore determined by precisely the objects that appear in them. To differentiate fields of sense from one another, we need to have a sense of their meaning. To understand fields of sense, we therefore need other forms of knowing than ontology, including "experience, our senses, language, thought—in short, the reality of human knowledge."[22] Fields of sense, in other words, cannot be fully understood ontologically, but only experientially or empirically.

Since Gabriel submits, "reality *itself* is infinitely complex," i.e., since the "combinations" of objects are infinite, the ways in which we perceive them are infinite as well. This means that there is an "infinity of fields of sense," i.e, ways of viewing things.[23] Importantly, we are not necessarily the "authors" of fields of sense: not every field of sense is the result of human activity, as is, for instance, the case with galaxies or elementary particles. "Brute facts of reality that we can do nothing to alter" exist in the same fields of sense in which we do. What this means, he suggests, is that we "cannot sort reality into the mind on one side and the world on the other" because the mind is "actively involved in reality."[24] We cannot, therefore, posit some sort of "'extra-mental' reality" as a measure to determine what it "means to exist and to be real."[25]

According to new aesthetic realism—which Gabriel asserts to be a "global trend in contemporary philosophy"—there is therefore no such thing as an external reality, there is no world "out there," as Kant postulates, "without spectators," nor can we consider "ourselves as beings outside of perceptible reality."[26] Consequently, like Harman, Gabriel rejects the phenomenological model of perception (subject vs. object, self vs. world) because it suggests that we can never perceive things as they are, since "our only way of representing [something] obscures it from our perception by highlighting at most a point of

---

20 Markus Gabriel, *Why the World Does Not Exist* (Cambridge UK: Polity Press, 2018) 232. Hereafter *WW*.

21 Gabriel, *Why the World*, 87.

22 Gabriel, *Why the World*, 87.

23 Gabriel, *Power of Art*, 25.

24 Gabriel, *Power of Art*, 25.

25 Gabriel, *Power of Art*, 26.

26 Gabriel, *Power of Art*, 26.

view of the object."[27] In relation to art, this means that art is never truly part of reality since phenomenological skepticism denies the very possibility of our capacity to "perceive things as they are." This, Gabriel claims, is "profoundly wrong" since "we can in fact perceive objects and know many things about them thanks to perception."[28] This is not to say that Gabriel equates our perception of a voice, say, or a chair, with our perception of works of art. What distinguishes works of art from other objects of perception is that they "enter into the relation of perception" by way of our relation to them.[29] Gabriel thus posits a three-way relation of perception, which "holds" between ourselves and an object, but which also "adds another dimension to reality," which is the "form of the relation."[30] It is this three-dimensional model of perception that centrally supports his overall claim to the structural, ontological power of art.

Let me backtrack a little to explain the general supposition underlying this new aesthetic realism. Whereas objects, we have seen, assume a critical place in Gabriel's thought, he forthrightly proclaims that the fact that we can "come up with theories of the universe" sets us humans apart from other animals.[31] This allows him to reserve a central space in reality for the imagination. Imagination, he writes, is "not a nothingness" nor does it "transcend reality in any way"—on the contrary, since what we imagine is real "to the extent that we imagine it," imagination is a "compartment" of reality.[32] This is crucial for a philosophy of art, since the "way in which we receive art is never purely perceptual."[33] Aesthetic experiences activated by art, and the "dreams" and "reveries" it evokes are therefore as real as it gets, since they add, rather than take anything away from reality. And whatever takes hold of our imagination, Gabriel boldly proclaims, has "absolute power over us."[34] The power of art on our aesthetic experience cannot be explained by the Kantian tradition of thought, which, in this respect, is sustained by OOO, because it is based on the idea that the artwork (or real object) withdraws from our perception. Since artworks are, in fact, found in reality, and since "human perception does not look onto reality from elsewhere," it is precisely the co-presence of work and

---

27 Gabriel, *Power of Art*, 27.

28 Gabriel, *Power of Art*, 28.

29 Gabriel, *Power of Art*, 29.

30 Gabriel, *Power of Art*, 39.

31 Gabriel, *Power of Art*, 8.

32 Gabriel, *Power of Art*, 10.

33 Gabriel, *Power of Art*, 34.

34 Gabriel, *Power of Art*, 9.

beholder in the same domain, or field of sense, that allows for aesthetic experience in the first place.

To be sure, and as suggested, we do not perceive artworks in the same way that we do ordinary objects. As ideas expressed in material, artworks form a combination of three elements: the material, its shape and the idea expressed. Importantly, however, artworks are "organized by an idea which unites a diversity of sensory elements."[35] As such, they are unequivocally different from, say, a table or my doorbell, to which I stand in a different perceptual relation. Art makes me feel as much as it makes me think—thought that is not necessarily conceptual—and what I perceive and what I think is not external to the aesthetic experience: these are activities that are "part of the artwork itself."[36] To the extent that art makes us think, "unleash[] a stream of thoughts" in the beholder as soon as they enter into a perceptual relation with each other in overlapping fields of sense, the artwork is a realization of a composition that is itself non-material. Artworks, Gabriel observes, thus "differ categorically from ordinary objects ... in that they enter into the relation of perception in such a way that they perceive themselves."[37]. This somewhat paradoxical claim is based on the supposition that in the three-way perceptual relation, the object of perception has the "property of standing in the relation of being perceived," which is a "real event, not something imaginary arising in disembodied minds."[38] As compositions that are "organized by an idea which unites a diversity of sensory elements," the artwork "utilizes our nervous system and our mind in order to realize itself"—it is thus "by way of our aesthetic experience of them," that artworks "relate to themselves"within the intrinsic overlapping fields of sense that constitute the perceptual relation in the aesthetic act.[39] This relation is not "external" to reality but is "itself part of reality."[40] Recalling Fried's and Harman's invocation of the notion of theatricality in the aesthetic act, Gabriel defines this real event in terms of "performance," as a performative act that "essentially encompasses the relation of thinking between us" and the work.[41]

Perception, Gabriel submits, is "not 'in our heads'"—when I perceive an object in a space, an object that others can see as well—I do not make

---

35 Gabriel, *Power of Art*, 35.

36 Gabriel, *Power of Art*, 44.

37 Gabriel, *Power of Art*, 29.

38 Gabriel, *Power of Art*, 33.

39 Gabriel, *Power of Art*, 29.

40 Gabriel, *Power of Art*, 40.

41 Gabriel, *Power of Art*, 43.

something up in my mind, I actually perceive the object, which means that the "perception is really there." The fact that one individual cannot see the perception of another entails that the perception is a "relation between a given perceiver and her object" and that that relation is "not in itself an object one can perceive." In relation to *Threshold*, this means that we, as beholders of the photographs, do see the viewers in the gallery space looking at the (empty) walls, but what the artworks having been removed from the perceived walls conveys is precisely the fact that we cannot actually see a perceptual relation, which only takes place between a beholder and her object. In this way, the works in the series at once inscribe the reality of the perpetual relation and confirm that this relation, a perceptual one, is "not a further sense modality" but the "form assumed by the relation of perception."[42]

On the basis of his three-dimensional model of perception, Gabriel suggests that "artworks do not exist without being interpreted." The beholder is as crucial in his model of art as it is for Harman. However, rather than allowing the aesthetic object to disappear behind its sensory aspects and putting the beholder in its position, Gabriel argues that the "perception of art or our thinking about it" are not "external activities" but are "part of the artwork itself."[43] Interpretation, however, is not the same as theoretical analysis. Whereas some conceptual artworks provoke theoretical analysis as much as they invite interpretation, artworks generally do not need to be decoded "at the level of theoretical or conceptual analysis in order to transform them into artworks" by way of our own perpetual relation to them.[44]

Like Deleuze before him, Gabriel defines artworks as "compositions"—not the compound Harman has in mind when he talks about the work of art qua work, but the combination of material qualities (bronze, words, paint) and interpretations. Without the latter, artworks would be "mere objects of random artifacts." Interpretation is key since it connects us to the composition of the work. And to understand art in terms of composition leads us to the "autonomy of art." With autonomy, Gabriel here refers to the "property of something's being governed by its own law, of being subject only to those laws that constitute it as what it is."[45] Unlike human beings, who share a certain universal humanity, works of art are radically different from one another. Greig's *Gallery Horizons* may on some level represent a shared space in art galleries, the line where the floor meets the wall, but as the distinct titles of each of the photographs

[42] Gabriel *Power of Art*, 41.

[43] Gabriel, *Power of Art*, 44.

[44] Gabriel, *Power of Art*, 45.

[45] Gabriel, *Power of Art*, 47.

conveys, as works of art, these photographs differ from one another, fundamentally, in their individual composition. As a result of the radical difference between them, my perceptual relation with each photograph in the series is unlike any other, because there is "no content whatsoever that the composition of all artworks shares." Artworks, in Gabriel's terms, are "radical individuals, whereas human beings are individuals that partake of a universal form."[46] Artworks, in contrast, are "radically autonomous." The reason for this is that artworks combine layers of fields of sense: "Artworks are compositions of fields of sense."[47] As such, they operate from within their own internal law.

Unlike other fields of sense—such as a presidency or the "second arrondissement of Paris"—artworks depend on interpretation for their existence. A nation state is a field of sense, but it is subject to laws external to itself. The radical autonomy of artworks stems precisely from the fact that they are bound exclusively by their own laws. Even in terms of genre distinctions, there are no ultimate laws that determine what "genre" a certain work of art represents or belongs to. Any classification of art is an approximation that does not tell us much about the work that a given work of art performs in the world, since its power resides in its "pure singularity."[48] This leads Gabriel to the question of what art actually is, a question he responds to by saying the "art is the definition of art." Rather than an "empty tautology," it is this definition, which means that "art is identical to all the artworks," that allows for the radical autonomy of art: "*artworks are radically autonomous individuals*" [emphasis in the original].[49] Gabriel, in this context, refers to Duchamp's *Water Fountain*, which is not an artwork in itself but is turned into an artwork because of its composition in being brought into an exhibition—it is at this moment of composing that the *pissoir* becomes a radical individual; the artwork fundamentally depends on its composition. As such, it is brought into and constitutes a field of sense that elicits a performative participation of the beholder, which is neither stable nor pre-given.

Gabriel concludes from this that "artworks require artists for their being," but this does not mean that the artist has much to say about how we interpret their work. Whatever meaning the artist means to convey, this cannot be a "decisive norm for our interpretations."[50] A work of art can and will be interpreted differently depending on the specific field of sense it which it appears—

[46] Gabriel, *Power of Art*, 49-50.

[47] Gabriel, *Power of Art*, 51.

[48] Gabriel, *Power of Art*, 54.

[49] Gabriel, *Power of Art*, 55.

[50] Gabriel, *Power of Art*, 57

spatiotemporal shifts in interpretive engagement with the same work of art merely confirm that the meaning of a work of art does not exclusively reside in the work, much less in the intentions of the artist, but all depend on the fields of sense in which the work appears to a given beholder or viewer. Recall that "what individuates an artwork is its composition," which is "radically autonomous." Just as Harman claims that the work of the work of art is ultimately "unparaphraseable," so does Gabriel argue that the "composition of an artwork is not in itself completable."[51] As long as there is further aesthetic engagement with a work of art, the work will not be fully realized: as a compossible in a particular field of sense, its realization—unlike what Deleuze would call its actualization—will never be complete. This gives them, in Gabriel's terms, an "irreducible reality," which is a "feature"of their radical autonomy.[52]

Unlike natural phenomena that can be studied scientifically, artworks do not lend themselves to such analysis: their reality exists in the performative quality of interpretation. Theoretical analysis of art from whatever critical perspective is a "second-order investigation into the human mind's performance of artworks." This means that artworks are not a natural phenomenon: they exist first and foremost in and as events. Not unlike Deleuze's notion of the event, the artwork for Gabriel is a real encounter in and with a shared field of sense, in which the interpretive activity of the beholder is dynamic, something that happens, and which is hence processual. The event of the work of art can never be repeated twice: every new encounter with an artwork will not be predictable according to the laws of nature but will re-constitute both the work and myself in my performative (or theatrical) engagement with the work as such. Gabriel therefore posits that the radical autonomy of art "leads human life to an impasse." Since the radical autonomy of art means that "any perception or contact with it has to be part of the artwork,"[53] the event of the encounter entails that we, in this aesthetic experience, are "completely suck[ed] in the artwork." Rather than coming to stand in for the (disappearing) aesthetic object, as Harman would have it, I, as a beholder of the work of art —i.e., as an active beholder, who engages with the work's radical autonomy—become subject to its power. Existing in the same field of sense with a radically autonomous individual, I cannot escape the power of the artwork, and its force leaves me unable to leave, once I have entered it.

---

[51] Gabriel, *Power of Art*, 58.

[52] Gabriel, *Power of Art*, 68.

[53] Gabriel, *Power of Art*, 71.

All this does not mean that any odd encounter with an artwork has this same effect. One may or may not enter the artwork and "aesthetic experience [therefore] either take place, or it doesn't," Gabriel concedes. He comes very close to Deleuze's notion of the encounter, in which the artwork and the beholder co-constitute each other in a moment of the artwork's actualization. Gabriel's new realist aesthetic model renders the process of the encounter and the experience of the artwork qua event very concrete:

> To the extent to which I participate in the self-constitution of the artwork, the artwork performs itself on the stage of my mind. My mind transforms itself into a self-manifestation of art.[54]

This conceptualization of the aesthetic encounter means that the beholder is never looking from the outside in: rather, the beholder is kind of lost, or loses her/himself in precisely the moment when the artwork self-constitutes itself in the moment of the aesthetic encounter, the co-constitutive moment when the radically autonomous power of art overtakes the relative autonomous power of the beholder so that the work of art is not so much realized or completed but actualized in the performative process of active engagement and co-constitution.

Gabriel asserts that artworks "exist for no reason beyond themselves" (*PoA* 78). Human beings exist in many different capacities and hence are "autonomous agents insofar as we are able to unify our manifold appearances in different fields of sense." We are capable of self-reflection, conscious of the fact that we have self-consciousness, and we interact across a myriad of fields of sense, in which morality, politics, and laws intersect. Works of art do not: art, Gabriel concludes, is "*amoral, a-legal, and a-political. The power of art is therefore an absolute power*" [emphasis in the original] (*PoA* 80). Since artworks give themselves their own laws, they "stand in need of interpretation"—they exist for no reason and "just happen to be there" (*PoA* 82). Human thought also exists in and for itself, but the difference between artworks and human thought is that the former are a-logical. Since value, beauty, and truth exist in reality itself, the artwork does not disappear behind its own features. On the contrary, we can encounter and perform its values on the stage of our minds, only if "art does not ask us humans, whether or not we like it." In its radically autonomy, art exists in reality, and we may or may not decide to share its field of sense: "We are just drawn into artwork, or we are not. Such is the power of art" (PoA 88).

54 Gabriel, *Power of Art*, 73.

Fig. 5.5: Leslie Sacks, Alex Weinstein: *Alex Weinstein Paintings,* (Bergamot Station) archival pigment print mounted on Dibond,14.5 x 22 inches, 2014.

By positing the radical autonomy of art, by insisting on its existence in reality, and in a potentially shared field of sense with the beholder, Gabriel's new aesthetic realist model resolves two problems I encountered when trying to think about Greig's work in the terms of OOO's theory of aesthetics. First, by positing the aesthetic act in the context of a three-way perceptual relation in intrinsically overlapping fields of sense, it avoids reducing artworks to objects, which turns them into "mysterious and unintelligible forces" whose essence escapes us.[55] This runs counter to the fact that art does things in the world, is found in a reality which it not only is capable of enriching, but in which it also has been and continues to be considered a dangerous force to reckon with: the long history of book burnings, and the destruction of sculptures, paintings, and other forms of "inappropriate art" is merely one fact that negatively attests to this truth. Second, whereas Gabriel's claim that "artworks do not exist without interpretation," [56] that "[w]ithout interpretation, artworks would be mere objects or random artifacts" might appear to approach Harman's foregrounding of the beholder in the aesthetic act, the opposite is true. In OOO, we recall, the beholder comes to stand in for the object that withdraws.[57] In new aesthetic

[55] Gabriel, *Power of Art,* 46.

[56] Gabriel, *Power of Art,* 44.

[57] Gabriel, *Power of Art,* 47.

realism, in contrast, the object—in this case both the material art-object and its non-material composition in perceptual relation with the viewer—not only does not disappear, but instead exists in reality, while also adding to this reality by operating in this intrinsically rather than in an external relation to it.

As singular "compositions of fields of sense," artworks are "essentially interpretation-dependent,"[58] Gabriel argues, but such "interpretations cannot be anticipated or controlled" by the artist, nor is the composition of an artwork ultimately "completeable."[59] As long as there is aesthetic engagement with an artwork, it continues to be realized, gives rise to new or changing feelings and thoughts, and therewith continues to think itself. At the same time, there is "no art without sensibility," for artworks "rely on realities we can only perceive."[60] *Gallery Horizons* confuses and problematizes the relation between what I perceive (the photograph as surface, its subject—the meeting point between floor and wall, and the imaginary realities that are generated in the perceptual illusion of a landscape that realizes itself in my thoughts and feelings) in the actual aesthetic experience. Recall furthermore that my first encounter with the work Matthew Marks/*Brice Marden: New Paintings* immediately called to mind an early perception of the relation between sky and the sea, between airspace and ground, thus adding yet another layer to the perceptual reality of the floor/wall-deep landscape composition. Instead of becoming inscrutable behind its sensory qualities, as does the real object in Harman's account of the aesthetic act, the form of the perceptual relation, as proposed by Gabriel, adds another dimension to reality, enriches it, by creating layers of actual and imaginary perception that do not exceed or transcend the object of perception but rather converge in a composition that is both real and concrete, intrinsic to the reality in which the artwork and I co-exist, in a shared field of sense.

---

[58] Gabriel, *Power of Art*, 51.

[59] Gabriel, *Power of Art*, 58.

[60] Gabriel, *Power of Art*, 67.

Fig. 5.6: Craig Krull, Ned Evans: *New Paintings and Resin Reliefs, Paintings* (Bergamot Station) archival pigment print mounted on Dibond,14.5 x 22 inches, 2014.

Rather than erasing the so-called trace of the real or turning the perceptual world into a nihilist simulacrum, the digital art photographs that provoked these thoughts thus give rise to an aesthetic experience from which the real object has not so much disappeared, and in which I come to stand in for the object that has withdrawn, but in which the three elements of the artistic composition converge. As suggested, the artwork performs itself, as Gabriel puts it, "on the stage of my mind."[61] By participating in it, the artwork seizes hold of me, I become part of it, and I cannot escape it. By situating these sites of perception in actual art spaces, *Gallery Horizons* at once reinscribes the function of such spaces for concrete and singular practices of looking and transforms them from within, thus adding yet another layer to an already multiple perceptual reality. Such, then, is the work of the artwork in its incompleteability generally. But perhaps, more specially, perhaps also the work of the digital photograph that shows us something about photography per se, whether analog or digital, i.e., that it was never, to recall Sigrid Nuñez's words with which I opened, to either preserve or destroy the past, but to add to our present reality in all of its infinite complexity.

[61] Gabriel, *Power of Art*, 73.

Bibliography

Gabriel, Markus. *Why the World Does Not Exist.* Cambridge, UK: Polity Press, 2018.

Gabriel, Markus. *The Power of Art.* Cambridge UK: Polity Press, 2020.

Greig, Cynthia. *Gallery Horizons* (artist statement), 2018. n.p. https://www.cynthiagreig.com/cynthiagreig/art/galleryhorizons/ghindex.html.

Harman, Graham. *Art + Objects.* Cambridge, UK: Polity Press, 2020.

Nunez, Sigfrid. *The Friend: A Novel.* New York: Riverhead Books, 2018.

Chapter 6
Connecting Modernism and New Media Theory: Bruno... Pound. Flusser.. Joyce

Aaron Jaffe
Florida State University

Abstract

This chapter tracks the intersection of modernist studies and new media theory, focusing on Flusser's concept of the technical image. It traces the intellectual lineage of this concept through figures like Giordano Bruno, Ezra Pound, and James Joyce, highlighting their shared interest in a decentered, infinite universe and the role of the metamedium in shaping perception and critical practice. The analysis examines how these modernists, viewed as technologists of the risk-medium, challenged traditional notions of form and meaning, embracing risk and incoherence. The chapter further connects this modernist approach to Flusser's later work, emphasizing the significance of medium orientation in a world saturated with technical images. The chapter uses Samuel Beckett's chapter title, "Dante... Bruno. Vico.. Joyce," as a framework to structure this intellectual genealogy.

Keywords: Technical Image, Modernism, Giordano Bruno, Media Theory, Vilém Flusser, Ezra Pound, James Joyce, Ideogram, Cosmology, Risk.

***

> Instead of representing the world, [technical images] obscure it until human beings 'lives finally become a function of the images they create. Human beings cease to decode the images and instead project them, still encoded, into the world 'out there,' which meanwhile itself becomes like an image — a context of scenes, of states of things.
>
> In the world of apparatus, all "waves" are made up of grains, and all "processes" are made up of punctuated situations.
>
> –Vilém Flusser[1]

[1] Vilém Flusser, *Towards a Philosophy of Photography London* (London: Reaction, 2000), 10, 67.

This chapter seeks to open a channel - solve a puzzle; repair a frayed wire, connect the dots, as it were - between modernist studies and new media theory, specifically focusing on the genesis of the concept of the technical image, as employed by the godfather of media theory, Vilém Flusser. When the German media theorist Friedrich Kittler writes "New media do not make old media obsolete; they assign them other places in the system," he is alluding to preoccupations derived from Flusser, who Kittler invited to lead his university seminar in Bochum right before Flusser's death in 1991.[2] For Flusser, technical images - artifacts produced by and for tech - assign words and images new, harder to reach places in an inhuman, infographical cosmology.

With the appearance of the technical image, he writes,

> the orderly threads finally fall apart, and the concepts lose coherence. In fact, the situation disintegrates into a swarm of particles and quanta, and the writing subject into a swarm of bits and bytes, moments of decision, and molecules of action. What remains are particles without dimension that can be neither grasped nor represented nor understood. They are inaccessible to hands, eyes, or fingers. But they can be calculated (*calculus,* "pebbles") and can, by means of special apparatuses equipped with keys, be computed. The gesture of tapping with the fingertips on the keys of an apparatus can be called "calculate and compute." It makes mosaic-like combinations of particles possible, technical images, a computed universe in which particles are assembled into visible images. This emerging universe, this dimensionless, imagined universe of technical images, is meant to render our circumstances conceivable, representable, and comprehensible.[3]

For Flusser, technical images, compiled from discrete states, ones and zeros, are not surfaces, but "mosaics assembled from particles."[4]

---

[2] Friedrich Kittler, "The History of Communication Media," *Ctheory* (1996), https://journals.uvic.ca/index.php/ctheory/article/view/14325/5101. See Daniel Raschke, "Flusser on the Fly: Towards an English Translation of Flusser's Bochum Lectures," *Flusser Studies* 33, https://www.flusserstudies.net/sites/www.flusserstudies.net/files/media/attachments/raschke-flusser-on-the-fly.pdf; and, Aaron Jaffe, The Double Repatriations of Cybernetics and Media Theory: Vilém Flusser and Heinz von Foerster, *New Literary History* 44, no. 2 (Spring 2023).

[3] Vilém Flusser, *Into the Universe of Technical Images*. Translated by Nancy Ann Roth. Minneapolis: University of Minnesota Press, 2011, 10.

[4] *Ibid.*, 6.

Modernity gets systematically reoriented to a hidden administrative calculus; a punctuated situation digitally transformed into ubiquitous signals; communications, subject to second thoughts, risky calculations and various analogical suspicions about metadata. There is, in all this, a non-trivial connection between Flusser's ideas and a certain high modernist bookshelf. The reading list follows above all from a reappropriation and repurposing of the concept of the ideogram during the first decades of the 20th century.[5] The ideogram is referenced in the work of Walter Benjamin and Sergei Eisenstein, but Flusser specifically gets it from Marshall McLuhan and Hugh Kenner, who in turn crib it from the controversial translation of the sinologist Ernest Fenollosa by Ezra Pound, called *The Chinese Written Character as a Medium for Poetry* (1919).[6]

This reference and others are found in the books of Flusser's travelling library - currently held by *Akademie der Künste* in Berlin. Flusser's actual bookshelves store all the relevant receipts, modernist annotations linking an expanded, media-theoretical hermeneutics not just to the ideogrammatic method, underscored emphatically in his copy of *The Literary Essays of Ezra Pound*, but also to the leads to analogous media cosmologies invoked in my title, namely, James Joyce, Giordano Bruno, Samuel Beckett, among others, interesting in this context precisely because they comprise the deep background noise of media-modernity.[7] To state my thesis as plainly as possible: Vilém Flusser derived his highly influential account of the technical image from the media dreamwork contained in the modernist sources on his book shelves.

My subtitle above alludes to Samuel Becket's very first critical essay - "Dante… Bruno. Vico.. Joyce" - published in 1929 in *transition*, the avant-garde magazine.[8] Appearing seven years after Joyce's *Ulysses*, the piece was partly strategic, partly a PR stunt to kickstart a reception horizon for *Finnegans Wake*, a work which eventually fully materialized ten years later. Whereas Dante and

---

[5] See esp., Seth Morton, *Modernist Interference: Inhuman Frequencies Between Modernism and Media* (Ph.D. diss. Rice University, 2019).

[6] Evan Kindley, "Kenner's Networks," *Critical Inquiry* 47 (3):525-543 (2021). See esp., Guy Stevenson, "Fascism, Iconoclasm, and the Global Village," Daniel Irrgang, "Vilém Flusser in Open Circuits," and Daniel Raschke, "Flusser and Ars Electronica: Between and Beyond Cybernetics," *Understanding Flusser, Understanding Modernism* New York: Bloomsbury, 2022). Ernest Fenollosa and Ezra Pound, *The Chinese Written Character as a Medium for Poetry: A Critical Edition*, Edited by Haun Saussy, Jonathan Stalling, and Lucas Klein (New York: Fordham University Press, 2008). Hugh Kenner, *The Pound Era* (Berkeley, CA: University of California P, 1971), 160, 275-276.

[7] Ezra Pound, *The Literary Essays of Ezra Pound* (New York: New Directions, 1968).

[8] Samuel Beckett, "Dante… Bruno. Vico.. Joyce," *Disjecta: Miscellaneous Writings and a Dramatic Fragment* (London: John Calder, 1985), 19-35.

Vico have been repeatedly name-checked as decisive modernist influences, allusions, precursors and heroes, Giordano Bruno (1548-1600), the modern heretic, thinker, and cosmologist, may be more surprising and less obviously comprehended. Joyceans certainly picked up on the reference and resonance of the Nolan - one of the Bruno's many appellations - and his various heretical positions across the Dubliner's oeuvre. Bruno also plays a critical role in Pound's corpus from his early writings in the *New Age* to the final *Draft and Fragments* section of *The Cantos* as a protagonist for art-as-free-speech amidst the diversity of matter. For Pound, like Joyce, Bruno provides an important avatar for theorizing modern aesthetic revolution in a media-cosmological sense, specifically, grappling with notions of incoherence and heteronomous infinity. Even more surprisingly, perhaps, the reference to the Nolan is decisive for Flusser directly, who invokes Copernicus, Bruno, and Galileo as a kind of unholy trinity of medial materialism, underscoring perspective, orientation, and risky relativism. Inspired by Joyce and Pound, re-read through the lens of cybernetics and his own descriptive and interpretive phenomenology, Flusser's Bruno is a decisive thinker of media orientations in the heteronomous, decentered universe of technical images. Tracing this genealogy through Pound, Flusser, and Joyce, my proposed infograph charts the persistence of Bruno-ism as ambient background noise in an expanded genealogy of modernism's medialects and *cultura experimentalis* (to borrow Siegfried Zielinski's phrase).[9] As Beckett's title intimates, connecting the dots is important to the understanding the long game of tracking technical images across punctuated situations into the modernist telematic mediaverse.

One of the reasons Beckett's essay is extremely difficult to track is its performative sleight of hand about negations, refused connections, contraries that trade on ironic rebounds. The most famous articulation of this concept comes from an essay by Fyodor Dostoevsky: "Try to pose for yourself this task: not to think of a polar bear, and you will see that the cursed thing will come to mind every minute."[10] In terms of *not* thinking of images of polar bears, Beckett may claim to caution against risky identifications, analogy-mongering, and gestural coincidences, but these kinds of warnings actually set the boomerang effects in motion. *Philosophy and philology are ... carefully folded like ham sandwiches. Literary criticism is ... a kind of book-keeping.*[11] Neatness, care, completeness, pigeon-holing, and abstractions have secret ministries that turn tidy jobs into exquisite messes. Indeed, when Beckett questions if "must we wring the neck

[9] Siegfried Zielinski, *After the Media* (Minneapolis, MN: U of Minnesota P, 2013), 125ff.

[10] Dostoyevsky, "An Essay Concerning the Bourgeois," *Winter Notes on Summer Impressions* (Evanston, IL: Northwestern UP, 1997), 49.

[11] Beckett, 19.

of a certain system in order to stuff it into a contemporary pigeon-hole" in fact names precisely what he is doing in the essay - or, what he claims to observe happening in the world of Mr. Joyce's "Work in Progress": detecting signals that punctuate intolerable complexities: a "mountain, the coincidence of contraries, the inevitability of cyclic evolution, a system of Poetics, and the prospect of self-extension," to cite Beckett's words.[12]

For what it's worth, Beckett shared drafts with Joyce, an event referenced in the first letter collected in *The Letters of Samuel Beckett*, to the extent that some scholars credit Joyce with a hand in helping to compose it.[13] Then, there's the issue of the cryptically punctuated title: "Dante... Bruno. Vico.. Joyce." Three dots, one dot, two dots is deliberately difficult to parse. Is it pattern - or, ellipsis, the linguistic *et cetera*, the 27th letter, evocative of the mathematical complexities of an infinite, non-repeating string? In Morse code, dot-dot-dot, dot, dot-dot, that is S-E-I. The German word SEI is the second-person singular imperative form for SEIN (BEING). *Beckett*, in other words, communicates a signal: *let it be... Sei ruhig. Bleibe ruhig.. mein Kind. Be quiet... stay quiet. my child..*, in the words of Johann Wolfgang von Goethe.[14] In one of the many moments in *Wake*, when Beckett-*fils* appears, Joyce underscores a connection, the very sort of jokey bit Joyce makes regularly: *We drames our dreams tell Bappy returns. And Sein annews. We will not say it shall not be, this passing of order and order' s coming.*[15] Indeed, Joyce references Morse code in the same way, involving the names of two Sams: Beckett and Morse. The idea that Beckett and Joyce are playing with dots and dashes - plying the medium with stabs and gashes - isn't too farfetched, I think. More to the point, if S.E.I. names a relation to BEING - Seiendes means ENTITIES - *beings* precisely the way that Beckett means when talking about Joyce's writing. It shouldn't be read as about something but as a technical form factor. It is that something itself: the punctuation mark - the digital stab between the gashes - the contrary in an infinite continuum.

The common gloss on Beckett's title is that it roughly punctuates a temporal relationship between four names. The stab between each gash designates rough

---

[12] *Ibid.*

[13] To James Joyce, 23 March 1929: "Here is the latest insertion. I think it might follow the passage which treats of form as a concretion of content. I have succeeded in combining the three points in a more or less reasonable paragraph." Samuel Beckett, *The Letters of Samuel Beckett: 1929-1940* (Cambridge: Cambridge UP, 2009), 7.

[14] Johann Wolfgang von Goethe, "Der Erlkönig," *Selected Poems* (New York: Penguin, 2005), 44.

[15] Joyce, *Finnegans Wake* (New York: Viking, 1966).

digital units of time: "The periods in the title represent centuries between the authors listed."[16] The missing name is probably Beckett, scholars sometimes speculate. Three centuries between Dante (1265-1321) and Bruno (1548-1600). One beat between Bruno and Vico (1668-1744). Two more beats get us to Joyce (1882-1941). If it's a repeating pattern, we could surmise three more beats between the unnamed Galileo (1564-1642) and unnamable Beckett himself (1906-1989). S-E-I is six in Italian, after all. The idea of many readers that Beckett is producing a Joyce-sanctioned lineage of names, a genealogy of influences or precursors, a model of descent, or a receipt of generational lineage, but this couldn't be more ill-conceived.[17] If not heroic authorial role-models *per se*, the names present instead a roll-call of technologists of the medium.

Considering the roster, let's pass over Dante quickly, only because his importance for Joyce, for Beckett in particular and for modernism more generally is well established. From a media-theoretical perspective, Dante is, simply put, the originator of vernacular as a modern literary medium. At the end of *Portrait*, for example, Stephen writes in his diary that he "Turned off that valve at once and opened the spiritual-heroic refrigerating apparatus, invented and patented in all countries by Dante Alighieri."[18] *Literary criticism* may *not be bookkeeping*, as Beckett puts it, settling accounts, but Stephen reminds us that the diary is an apparatus for administering modernity, a symbolic form-factor that transforms selfhood into a continuously updating database, a register of biopolitical metadata. And, in this connection, he offers Dante as a technical add-on, a libidinal cooling machine. Transmuting physical states of matter like a swamp cooler, providing a different kind of Blast-valve for modernist innovation: the chill-out, the red shift. If *Ulysses* modernized Homer, it happened through marvelous feats of miniaturization and signal compression, the negative red shift: ten years on byways of the Mediterranean squeezed into eighteen hours in the streets of Dublin. Young Joyce also told his friend Padraic Colum as much,

[16] Washington University Image Collections and Exhibitions, http://omeka.wustl.edu/omeka/items/show/14843.

[17] Eyal Amiran, "Apparent Offspring: Joyce's Beckett's Succession in the 'Wake'," *The Centennial Review* 37, no. 2 (Spring 1993): 401-414: "three chosen precursors, and only begs for Beckett's name to be appended at the end. Each period in the essay's title corresponds with a rough century and hints at the model of literary descent that Joyce considered adequate to himself, one of a generational lineage of neoclassical authors, each turned towards an occult relation with the divine."

[18] Joyce, *Portrait of the Artist as a Young Man* (New York, Huebsch, 1916), 298. Howard Helsinger, "Joyce and Dante," *ELH* 35, no. 4 (December 1968): 593.

that Dante's *Divine Comedy* was Europe's epic paradigm – a view not incidentally borrowed from Vico's *La Scienza Nuova*, the *New Science*.[19]

Furthermore, as every Joycean knows, Vico's treatise is often said to provide the theoretical apparatus - or, scaffolding - for the *Finnegans Wake*. Yet, in Beckett's essay, Vico's relevance isn't about reparative mysticism via a method for decoding mythic archetypes or epiphanic fizzle but rather as a real-life semiotic-machine, an actual input/output algorithm which may or may not resemble, as Beckett says, nothing besides "a carefully folded ham-sandwich," which layers philosophical abstractions, philological concretions, and plenty of sandwich creme. Remember the Joycean word - the revolution of the word, in effect, is "not about something; it is that something itself," as Beckett famously puts it in the essay.[20] In effect, here he underscores the primacy of language as a technical medium.

The version of Vico appropriate to Joyce, Beckett explains, isn't about conjuring a mystic or a scientific guru - a role-model, in the old sense. Chiefly, Joyce matters as an innovator - a technologist - of the medium. The connection to innovation - to a new science of the medium - comes because his program, like Vico's, represents "an impertinent outrage against tradition."[21] Think of the convolutions of Viconian history not as a progression, a Dantescan stairway to heaven, but as an input/output algorithm of risk positioning, moving back and forth through stations of "necessity, utility, convenience, pleasure, luxury, abuse of luxury."[22] The final upgrade involves ambivalent results – heresies, in other words, like laziness, enabled by technical contrivances and machinic conveniences: "Mr. Joyce's Terrestrial Paradise [has] the tradesmen's entrance on to the sea-

---

[19] Richard Ellmann, *The Letters of James Joyce*; See also, Donald Phillip Verene, *James Joyce and the Philosophers at Finnegans Wake* (Evanston, IL: Northwestern University Press, 2016) and *Vico and Joyce* (Albany, NY: State University of New York Press, 1987. Helsinger, 592: "Unheard of before or since, the apparatus is a puzzle, unknown to cryologists, unrecognized by Dantists, and ignored by Joyceans. Whatever the device may be, in Stephen's conversation it operates at the turn of a valve to avert the personal and the painful. The flood of emotionalism thus stanched is kin to half the waters that whelm through the book, waters which are at once waters of life and yet to Stephen unpleasant. The incident is an invocation of Dante to protect the would-be exile from the lure of a worldly Beatrice. Joyce's idea of exile and his thoughts on the relationship of the artist to reality are thus involved with his opinion of Dante. In examining these ideas, we shall come to understand the nature of the spiritual heroic refrigerating apparatus.

[20] Beckett, "Dante... Bruno. Vico.. Joyce," 27.

[21] *Ibid.*, 20. See Robert Baines, *Philosophical Allusions in James Joyce's Finnegans Wake* (Oxford, UK: Oxford University Press, 2023).

[22] Beckett, "Dante... Bruno. Vico.. Joyce," 21.

shore," Beckett says.[23] Not a conical ascent upward, but rods and cones inside spheres: spyglasses, binoculars, microscopes, magic lanterns. The big picture, the discovery of a risky medium, comes first… and last, with diverse ingredients strewn about and sandwiched between. *His father told him that story: his father looked at him through a glass: he had a hairy face.*[24] Bruno's part in the recipe concerns orientation in an infinitely vast, center-less universe full of diverse materials.[25] Once a Dominican monk, Bruno raised the stakes of Copernicus's revolutionary heliocentrism to the level of heresy. He even borrowed from Aristotle for his reasoning. Aristotle, Bruno states, said that there is no absolute location for a body but everything corresponds to other things. Bruno's risky proposition was that, not only was the Earth <u>not</u> the center of a closed system, but everything moves and changes, so that a *middle* is a technical effect of observation. The sun is a star, with diverse exoplanets; an infinitely and diverse universe, full of infinitely diverse galaxies, stars, exoplanets, assorted matter.

"No man, said the Nolan, can be a lover of the true or the good unless he abhors the multitude; and the artist, though he may employ the crowd, is very careful to isolate himself."[26] With these lines, Joyce begins his very first published essay "The Day of the Rabblement" written when he was 19. They convey precisely a strange cosmology of signal-tuning from Bruno that Beckett picks up, too. For Beckett, the takeaway from Bruno "proceeds from rather arbitrary data to philosophical abstraction":

> There is no difference … between the smallest possible chord and the smallest possible arc, no difference between the infinite circle and the straight line. The maxima and minima of particular contraries are one and indifferent. Minimal heat equals minimal cold. Consequently, transmutations are circular. The principle (minimum) of one contrary takes its movement from the principle (maximum) of another. Therefore, not only do the minima coincide with the minima, the maxima with the maxima, but the minima with the maxima in the succession of transmutations. Maximal speed is a state of rest. The maximum of corruption and the minimum of generation are identical: in principle, corruption is generation.[27]

---

[23] *Ibid.*, 32.

[24] Joyce, *Portrait*, 1.

[25] Giordano Bruno, *On the Infinite Universe and Worlds.* New York: Henry Schumann, 1950.

[26] Joyce, "The Day of the Rabblement," *The Critical Writings of James Joyce*, ed. Ellsworth Mason & Richard Ellmann (NY: Viking Press 1959), 68.

[27] Beckett, "Dante… Bruno. Vico.. Joyce," 21.

Simply put, modernist media technologists are heretics of form. For Beckett's Joyce, even the world-form is too mundane to fit the vast plenitude of data. Remember Dedalus and his trouble with names and addresses: "He opened the geography to study the lesson; but he could not learn the names of places in America. Still, they were all different places that had different names. They were all in different countries, and the countries were in continents and the continents were in the world and the world was in the universe."[28] Attempting to generate biographical exhaust, Stephen runs through and out of relevant metadata. Tellingly, young Dedalus scales this ladder quite specifically into nested gestures of record-keeping for what Bruno calls "innumerable worlds": from the closed world of textbook page ("himself, his name and where he was") to the radical outdoors of the infinite universe, echoing the title of the magisterial work by the French-Russian philosopher of science Alexandre Koyré.[29] *What was after the universe?* he asks: "Nothing. But was there anything round the universe to show where it stopped before the nothing place began? It could not be a wall but there could be a thin thin line there all round everything. It was very big to think about everything and everywhere."[30] Returning to Beckett's account of Joyce's cosmology: the point is that Stephen is wrong, because the nothing is not nothing.

The relevant, media-theoretical point is, instead, that the universe is horizonless; there is no thin, thin line that limits, a conclusion leading Bruno not to despair but to a kind of frenzied enthusiasm. Facing imminent execution, he drags a fellow prisoner "to the window and showed him a star, saying that it was a world and that all the stars were worlds."[31] Bruno, Koyré recounts, with some irony, presented this new cosmology "with a burning enthusiasm that of a prisoner who sees the wall of his jail crumble that he announces the bursting of the spheres that separated us from the wide open pace and inexhaustible treasures of the everchanging, eternal and infinite universe."[32] Perhaps his obliviousness about risks is the reason Koyré doesn't consider Bruno properly modern.

In linking Bruno to Vico, Beckett describes a duly modernist risk epistemology that punctuates contraries through the medium. I think this is the meaning of

---

[28] Joyce, *Portrait,* 11.

[29] Alexandre Koyré, *From the Closed World to the Infinite Universe* (Baltimore, MD: Johns Hopkins University Press, 1957).

[30] Joyce, *Portrait,* 12.

[31] Alberto A. Martínez, "Was Giordano Bruno Burned at the Stake for Believing in Exoplanets?" *Scientific American,* March 19, 2018, https://blogs.scientificamerican.com/observations/was-giordano-bruno-burned-at-the-stake-for-believing-in-exoplanets.

[32] Koyré, 43.

the punctuated names in his essay's title. The telematic medium, as he puts it, is "neither to be considered as a formless structure, due exclusively to the achievements of individual agents, nor as possessing reality apart from and independent of them, accomplished behind their backs in spite of them, the work of some superior force, variously known as Fate, Chance, Fortune, God."[33] For the second modernists, Beckett indicates, the instrument is thing, the machine: "The lamp is more important than the lamp-lighter."[34]

> The vicious circle of humanity is being achieved, and this achievement depends on the recurrent predomination of one of two broad qualities. No resistance, no eruption, and it is only in Hell and Paradise that there are no eruptions, that there can be none, need be none. On this earth that is Purgatory, Vice and Virtue - which you may take to mean any pair of large contrary human factors - must in turn be purged down to spirits of rebelliousness. Then the dominant crust of the Vicious or Virtuous sets, resistance is provided, the explosion duly takes place and the machine proceeds. And no more than this; neither prize nor penalty; simply a series of stimulants to enable the kitten to catch its tail. And the partially purgatorial agent? The partially purged.[35]

Prize or penalty, vice or virtue, risk or consequences, paradiso or apocalypse: the closed inner space of the living, like ones and zeros in a cosmic double-entry account book - is framed by the possibility of an infinitely risky, inhuman outer-verse: the image of a recording angel (Beckett/Joyce) is seconded, doubled, by the image of his master's voice (Joyce/Beckett).

As I have mentioned, the misplaced middle in Beckett's strange four-fold might as well be Galileo Galilei. Galileo actually gets name-checked in Beckett's poem *Whoroscope*, published in 1930 by Nancy Cunard's Hours Press.[36] The *et cetera* effect implied by the non-repeating infinite string of his *Exagmination* essay is theoretically limitless: the travelling library of Bruno of Nola; Beckett himself; Cusa, Nicholas of; Copernicus; Flaubert; and on and on and on. Ostensibly, Beckett's poem is about Descartes and his repudiation of modern Copernican cosmology, inspired by incoming news through scholastic discourse networks of Galileo's own forced recantations. The cosmological implications, connected to Bruno's heretical reading of Copernicus, are the diversity of matter and trailing repercussions for the diversity of worlds and words. Free

---

33 Beckett, "Dante... Bruno. Vico.. Joyce," 21.

34 *Ibid.*, 22.

35 *Ibid.*, 33.

36 Beckett, "Whoroscope," *The Collected Works of Samuel Beckett* (New York: Grove, 1970), 11-17.

speech is very much the modernist cause *par excellence*, but it's worth mentioning that then as now it is not offered as an unalloyed good but as a value shot through with inescapable harms.

The pivot from modernism into medium of technical images is the persistence of the literary continuum of analogs in a universe of technical images - and it is found right there with the inquisition of Bruno. Now, the connection to Pound, I think, is a reel-to-reel equivalence between modern heresy and free speech that stretches across his oeuvre. It is strikingly prominent in the context of the frenzy of paranoia - paranoid modernism, to borrow David Trotter's phrase - in Pound's repugnant Radio Speeches for Mussolini. "Free speech without free radio speech is as zero."[37] When Pound presented himself to the American army in Genoa in 1945, these were his words: "If a man isn't willing to take some risk for his opinions, either his opinions are no good or he's no good."[38] Pisa's famous risk-taker Galileo was certainly on his mind. The reference to "Galileo index'd 1616" comes very early in Rock-Drill, written from incarceration at St. Elizabeth's, with a certain defiant undertone.[39] Pound states in Canto 85 that the "record of Confucius is the record of a very great sensibility."[40] From the grooves of "great sensibility," Pound zeros in on the new capacities for technical transfer of risky signals from medium to medium: "Galileo Galilei [who] discovers that the shadow of the moon bears a relation to the passage of time and that the movement of bodies is identical, in principle, to that revealed in the shadow of the pointer on a sundial (gnomon)."[41] When he writes in his Henry James essay that, "literature in the nineteenth and the beginning of the twentieth centuries is where science was in the days of Galileo and the Inquisition," Pound means to emphasize the risk-taking.[42] Kittler says something similar, invoking Rainer Maria Rilke and László Moholy-Nagy to show that "engineers and the avant-garde think alike" given enough time.[43] Technical transfer from medium to medium is a risky endeavor.

Pound, of course, was a strenuous advocate for the medium concept, in his case derived from Fenollosa's account of *The Chinese Written Character as a*

---

37 *Canto LXXIV.* See David Trotter, *Paranoid Modernism* (Oxford University Press, 200); Canto LXXIV.

38 Pound, *Selected Prose, 1909-1965* (New York: New Directions, 1975), 18.

39 Pound, *The Cantos of Ezra Pound.* London: Faber and Faber, 1975, 426.

40 *Ibid.*, 543.

41 *Ibid.*

42 Pound, "Henry James," *The Literary Essays of Ezra Pound* (London: Faber, 1954), 316.

43 Kittler, *Gramophone, Film, Typewriter* (Palo Alto, CA: Stanford, 1999), 46. Thanks to Ed Comentale-Dallis for this.

Medium for Poetry.[44] As Haun Saussy has observed, the inevitability of orientation to medium provides "the defining gesture of modernity in the several arts—releasing each medium from its servitude to content; here, making poetry out of language rather than out of ideas.... The Chinese written character is a medium, then (as the essay's title unobtrusively contends), and like any medium has always to be rescued from instrumentality."[45] The irreducibility of medium "reverse[s] the habit that makes a medium a mere go-between or pass-through and turns it into the true subject of articulation.... [It] would have to be neither vocal nor mimetic; it would serve neither speech nor representation." Pound's particular approach was highly influential.[46] Saussy cites Charles Bernstein on Louis Zukofsky: "Rather than making the language as transparent as possible, where these other [phonemic, graphic, morphological] qualities are repressed as a matter of technique . . . the movement is toward opacity/ denseness—visibility of language through the making translucent of the medium. . . . To make language opaque so that writing becomes more and more conscious of itself as world-generating, object-generating."[47] Fenollosa himself, in his original draft, used the plural mediums not medium in this key passage. The medium is media - and for modernity it comes sequenced first, not in the middle: "Failure or success in presenting any alien poetry in English must depend largely upon poetic workmanship in the chosen medium."[48] Notice that failure or success is shot through the medium. Even concerning the problem of translation - that modernism is *born* translated, as Rebecca Walkowitz puts it, moderns are born translated - born digital, born mediatized, etc. - is baked into a paradigmatically ambivalent medium for risk, a risk epistemology: Whether poor translation or transformative misreading comes down to an orientation to cultural field as an abstract space, a packed waiting room for subsequent risk triage.[49]

---

[44] Ernest Fenollosa and Ezra Pound, *The Chinese Written Character as a Medium for Poetry: A Critical Edition*, Edited by Haun Saussy, Jonathan Stalling, and Lucas Klein (New York: Fordham University Press, 2008).

[45] Haun Saussy, "Fenollosa Compounded: A Discrimination," *The Chinese Written Character as a Medium for Poetry: A Critical Edition.* (New York: Fordham University Press, 2008), 36.

[46] *Ibid*, 37.

[47] *Ibid*, 38.

[48] Fenollosa and Pound, 43.

[49] Walkowitz, *Born Translated: The Contemporary Novel in an Age of World Literature* (New York: Columbia UP, 2015). Three dominant received metaphors of cultural circulation dominate modernist studies today: translation, mediation, and worlding. They have much to recommend them. The semantic evocations of "translation," for instance, collapsing the transit of bodies and the translation of discourse, enables useful

The editors of the critical edition of the Pound/Fenollosa piece carefully line-up a sandwich of risk positions: "the essay separates its readers into two groups, those who care about poetry and those who care about the Chinese written character."[50] In the double entry, one side of the form yields risky words as faulty linguistics, the other, chancy poetic in-appropriations. Put simply, when it comes to risky modernity, the medium comes first. Medium is certainly a real abstraction - to turn to Marx - ready-made for economic discourse, the medium of exchange being that kept Pound in a particular state of perplexity. If we bracket hazardous words - not seeing through them, instead considering for a moment the invention of medium as *sine qua non* for modernity - this leads to a very different experience of Eliot's famous "Tradition and Individual Talent" essay. Eliot's point is not that the poet is an intermediary for expression but that

---

thinking about the place of material embodiment in modernist discourse. "Mediation" has strengths, too, purposing inquiry into the relationship between modernist idiom and its material situation, infrastructure, and institutions. And, there is also something of this association in "worlding," the prevalent, generative metaphor for describing such phenomena with an eye on modernism's transnational ambitions. Yet, for all their utility, these terms conceptualize modernism's cultural movements in ways that are oddly stationary, uncharged, and motiveless. Instead, we favor the heuristic words of our title--coinage, circuitry, and switch-backs-terms which register, above all, that mapping the passages between modernist words and things entails a far more literal and rigorous understanding of "exchange," one which directly accounts for the regulative institutional, technological, and, yes, economic frameworks that make modernist culture possible. look more deliberately at modernism's distributional vectors: its metaphorics and mechanisms of circulation, exchange, and fungiblity. For all the very reasonable attention paid to the disruptive--departures, arrivals, end-stops and end-stations--modernist studies tend to overlook the complexities of the connective transit, transfer, interchange, flows, and currents. Charting the careers of modernist bodies and texts, tracking reading materials across desktops, cycling new ways of seeing through gallery and exhibition space: the lingering question is what conduits, metaphorics, homologies, elective affinities channel modernism's logic and development in, to, and through the cosmopolis. How are the pathways wired? Call it "mediation," "translation," or "worlding," modernism's investment in processes of investment, movement in, to, and through and matter--words, things, bodies, texts--is a problematic that merits more purposeful analysis. Such movement mirrors the ongoing transvaluation of objects in modernity, as capital flows apace in, to, and through multiple frontiers. Among other things, through a theoretical alertness to economic, lineal, and promotional traffic, we seek to move theoretical discourse about influence beyond its romantic sources into a new explanatory register more suited to modernist studies. Here, as elsewhere in modernism, causation is simultaneously uneven and over-determined, often multi directional, made possible by forms of influence rendered not with anxiety but as a kind of exposed circuitry.

[50] Saussy, 2.

expression itself as such is now off the table. Finely filiated platinum, as his famous analogy puts it, provides "a more finely perfected medium in which special, or very varied, feelings are at liberty to enter into new combinations."[51] The individual poet - there's that word, again - is "a particular medium, which is only a medium and not a personality, in which impressions and experiences combine in a peculiar and unexpected way."[52] What's discovered – or, newly understood – is medium as space, an analogical continuum, for new combinations of the peculiar and the unexpected. The medium concept pulls modernism/ modernity in several other directions: from the Greenbergian notion of modernist medium specificity and transparencies towards the opacities of a technical-medial a priori; from cybernetic environments towards mediation as interface effect with inhuman scalar possibilities. There is an acute sense that everything modern - every minute attempt to arrange our bodies and our lives - is about medium orientation. The future is reflexivity programming the medium that is the present in complex ways. The medium concept is slippery, if nothing else, slippery not least because of its kind of constitutive invisibility.

In Canto 48, following scenes of loudspeakers blasting official pronouncements from the Church, Pound puts modernity on blast "with the opinion (shouted under Pound's window) that Galileo was an imbecile who dumbed down the world."[53] What is up-for-grabs here is, I submit, orientation to a risky medium. *Galileo thinks he's special.* That response isn't wrong. He says it. From a certain, technically naïve orientation, one without a spyglass as it were - shared by many, including Aristotle - the message of the stars is an old one. Stars illustrate the upper limit of a closed, dusty world where everyone and everything has a place. With his telescope, Galileo deauthorized this bogus - yet commonsensical and somehow quite reasonable - doctrine. He fucked up the night. "In effect," as William Bysshe Stein notes in his reading of Beckett's "Whoroscope," Galileo "corroborated Descartes' basic premise" of "the unreliability of the senses in their apprehension of external reality": "when Galileo recanted at his trial for heresy, Descartes was dumbfounded,... '*if the movement of the earth is false, all the foundations of my philosophy are false.*'"[54] Much to his discredit, "always a trimmer in religious matters," Descartes devises absurdities to prop up corrupt

---

[51] Eliot, T.S., "Tradition and the Individual Talent," in *Selected Essays* (New York: Harcourt, Brace, 1950), 7.

[52] *Ibid.*

[53] Pound, *The Cantos of Ezra Pound.* London: Faber and Faber, 1975, 241. "Yours respectfully/Galileo Yours respectfully/Galileo, pronounced * Garry Yeo'/err 'un 'imbecille, ed ha imbecillito/(voice under my window) il mondo."

[54] Stein, 128.

universalisms endorsed by Church dogma, horoscopic hypocrisies derided by Beckett.[55]

The starry messenger is the starry medium itself. Pound's association with this discovery is almost without fail connected <u>not</u> with a constellation of names, but with the movement of starry messages - understood in a profounder sense as a proxy for free speech. And yet, he brings the name of Bruno back - a name Joyce sets in motion for modernism - in *Drafts and Fragments*, when he can't make any of it cohere. So, maybe a conclusion to draw is something like the constitutive heresy of open form, which is the constitutive environment/ system cybernetic distinction; the difference that makes a difference. Pound told Joyce that the first part of *the transatlantic review* would be reserved for him "with a trumpet blast" into the void. In another essay written in support of *Portrait*, Pound makes an explicit comparison between Galileo and Joyce based on their shared capacities to promulgate new messages:

> if one can't ... write ... as a physician, as a savant, as a historian, ... if we can't write plays, novels, poems or any other conceivable form of literature with the scientist's freedom and privilege, with at least the chance of at least the scientist's verity, then where in the world have we got to, and what is the use of anything, anything?[56]

Pound's support for Joyce's claim to a Nobel prize - a star-struck message he published in Berlin in 1924 - underscores two difficulties for Joyce's case. First, Pound records that Joyce, as an exile, like Conrad, gets necessarily excluded from rank and file nominations for national literatures and qualifies only as a member of "the republic of letters, or of the Heimatlos."[57] Those without papers. Second, he notes that Joyce's work had been deemed morally suspect for including too much:

> [T]here is one reason why no American or Englishman should be immediately eligible for any such international honour. Namely 500 copies of a great work of literature have been seized and burnt in Washington D.C. and another 500 seized and destroyed in Southampton; and no writer of standing resident in either England or America has emitted one word of protest: I refer to the seizure and destruction of copies of James Joyce's "Ulysses"; and if the Swedish Academy exists

[55] *Ibid.*

[56] Ezra Pound, *Pound/Joyce: the Letters of Ezra Pound to James Joyce* (New York: New Directions, 1970), 73. See Ian F. A. Bell, *Critic as Scientist* (London: Routledge, 1981).

[57] Pound, *Pound/Joyce*, 221.

> for the purpose of rendering some sort of international literary justice, there is this wrong to be balanced.
>
> You would think we were in the time of Bayle; when the Dictionnaire was burnt in market place.
>
> And at such a time it is fitting that a nation's authors should suffer for the ill deeds of the populace.[58]

The calculus is telling. Reversing the literary-political bureaucracies of censorship means inverting the popular charge that Joyce trades in the quotidian subject matter. Recall Virginia Woolf's dismissal, for example: Joyce was like a "a queasy undergraduate scratching his pimples."[59] *Ulysses* was" an illiterate, underbred book ... [by a] self-taught working man, & we all know how distressing they are, how egotistic, insistent, raw, striking, & ultimately nauseating...."[60] Pound counters such charges in a decidedly Brunoian way, *Ulysses,* he writes, manifests not base materialism but "the highest idealism," "that which tries to make others more aware of the relations between one thing, or one state of mind, and another; to make them more aware of the cosmos in which they exist."[61] High magic to low puns.

Pound's *Canto 114* is preoccupied by disoriented perspectives ...

Amid corridors and ambassadors

> glow worms and lanterns, and the moving is
> from the inward
> o di diversa natura (Giodorno Bruno)
> and in these triangular spaces[62]

And here, among the serious characters and forms, Pound once again finds Bruno, the cosmic geometer, pronouncing no difference between a rich infinite circuitry and straight lines. In a sense, Bruno is writing the distinctly modernist aesthetic playbook of fragmentation, medium primacy, discoveries of heteroclite materials and proximities to risky crises. It's all there in Giordano Bruno (1584): "We can assert with certainty that the universe is all center, or that the center of the universe is everywhere and its circumference nowhere.... There are no ends, boundaries, limits or walls which can defraud or deprive us of the infinite multitude of things":

---

[58] *Ibid.*

[59] Virginia Woolf, *The Diary of Virginia Woolf,* Volume II (New York: Harcourt, Brace, Jovanovich, 1980), 189.

[60] *Ibid.*

[61] Pound, *Pound/Joyce,* 220.

[62] Pound, *The Cantos,* 791.

> Make then your forecasts, my lords Astrologers, ... by means of those astrolabes with which you seek to discern the fantastic nine moving spheres; in these you finally imprison your own minds, so that you appear to me but as parrots in a cage, while I watch you dancing up and down, turning and hopping within those circles. We know that the Supreme Ruler cannot have a seat so narrow, so miserable a throne, so straight a tribunal, so scanty a court, so small and feeble a simulacrum that a phantasm can bring to birth, a dream shatter, a delusion restore, a chimera disperse, a calamity diminish, a misdeed abolish and a thought renew it again, so that indeed with a puff of air it were brimful and with a single gulp it were emptied.[63]

Bruno died for cosmology, writes Flusser.[64] He also died for modernism. And all this operational paranoia about signal detection and the modern medium - the impossibility of a symbolic form in the universe of technical images; coherence, in a word - happens during a long legitimacy crisis of literary humanism. We might call this situation second modernism.

Flusser's Bochum Lectures, undertaken at Kittler's invitation, betray a kind of comical self-awareness that the guise of literary-modernist refashioned into that of a media-theory guru is just more obligatory equipment for speakers telling things to post-literary audiences in academic settings.[65] When Flusser compares himself, not entirely ironically, to great sages on stages - Hegel in Berlin, Husserl in Göttingen, Bergson at the French Academy, Bachelard at the Sorbonne, later, emphatically, to Galileo in chains - one has a sense of hyperbole tinged by disappointed ambition. Flusser's late persona of media soothsayer is less reminiscent of an identifiable academic style and more evocative of wayfarers passing through the lecture circuit, as chiefly modeled by McLuhan. One might say further that the alien village that Flusser urgently sought to explain was above all a wired global village. Circuits, he tells us in the Bochum Lectures, are communicological formats, charmed circles, as it were, captive to motivated motifs and strange hierophantic enthusiasms. The idea of an intellectual as a networking hierophant on a lecture junket may itself be a hangover from an increasingly obsolescent age of writing, and Flusser's idea of communicology may be a feature, not a bug during this prolonged crisis of theory.

---

[63] *On the Infinite Universe and Worlds* (New York: Shuman, 1950), 245-6.

[64] Flusser, *Writings* (Minneapolis, MN: University of Minnesota Press, 2002), 194.

[65] Flusser, *Kommunikologie weiter denken: Die Bochumer Vorlesungen* (Frankfurt: Fischer, 2010). See Flusser, *Thinking Further: Fragments of Communicology*, ed. Aaron Jaffe et al., trans. Andrew Battaglia and Daniel Raschke, preface by Friedrich A. Kittler (Minneapolis: University of Minnesota Press, forthcoming 2026).

Yet, as expected, Flusser provides a remedy, an explanation of an alternate critical form factor, namely, the prominent persona - indeed, persona non grata - of the Big Man, an expression which Flusser unsurprisingly deploys in English. Neither media guru nor extraterrestrial amanuensis, this forlorn figure sits on a heap of technical detritus as a teller - a deep-time critical story-teller, using the heap as an apparatus for thinking about critical theory further. The somewhat torturous etymology and multilinguistic pun connect the verb to tell – *erzählen*, the Germanic cognates of account-keeping, recounting, critical narration and record-keeping – to the older, Hebraic remnants of the noun tell – in effect, this itself is the tell, the tip-off, underscoring a systems point about the expanding junkyard of technical dreams, the pile of information kernels and memories accumulating near where the grain is kept. On top of the middens, one sees further and tells more, Flusser says. Looking out, one sees what's coming, tells where the channels are and will be, reports how the risks of critical inflation and devaluation are already deeply programmed by the apparatus. In the Storyteller essay, Benjamin links the decline of critical frames to the degradation of experience. Benjamin writes: "A generation that had gone to school on a horse-drawn streetcar now stood under the open sky in a countryside in which nothing remained unchanged but the clouds, and beneath these clouds, in a field of forces of destructive torrents and explosions, was the tiny, fragile human life."[66] Flusser takes a similar view, but instead of a zombified husk framed by a horizon, he tells of the impossible heaps of broken, technical images that signal the ruin of the critical horizon itself. Flusser of the Bochum Lectures climbs the rubbish heap of tertiary retention, scans the waste data in garbage time, and thinks things further.[67]

Pound's relevance to the genesis of this strain of Flusserian media theory is worth noting. The link depends on his formulation of the image and the ideogram as a time-based "meta-medium" - to adopt Julian Murphet's word - "in which words and photographic images both became 'photographic' and 'printed' at the same time."[68] "The Chinese Written Character as a Medium for Poetry," for instance, highlights a certain kind of meta-medial technicity resonant with Flusser's concept of the technical image: "In what sense can verse, written in terms of visible hieroglyphics, be reckoned true poetry? It might seem that poetry, which, like music is a time art, weaving its unities out of successive impressions of sound, could with difficulty assimilate a verbal

66 Benjamin, "The Storyteller: Reflections on the Work of Nikolai Leskov," *Illuminations* (Schocken: New York, 1968), 83-4.

67 Flusser, *Kommunikologie weiter denken: Die Bochumer Vorlesungen.*

68 Julian Murphet, *Multimedia Modernism* (Cambridge: Cambridge UP, 2009), 63.

medium consisting largely of semi-pictorial appeals to the eye."[69] Reading backwards from Flusser to Pound (or, more accurately, to Pound's idiosyncratic adoption of Fenollosa) involves special attention to the ways media mediates spatial and temporal unevenness and conceptual simultaneity: epochal possibilities of media relations before alphabetic culture, to put it in a Flusserian framework.

Bibliography

Amiran, Eyal. "Apparent Offspring: Joyce's Beckett's Succession in the 'Wake.'" *The Centennial Review* 37, no. 2 (Spring 1993).

Baines, Robert. *Philosophical Allusions in James Joyce's Finnegans Wake.* Oxford, UK: Oxford University Press, 2023.

Beckett, Samuel. *The Letters of Samuel Beckett: 1929-1940* (Cambridge: Cambridge UP, 2009), 7.

---. "Dante… Bruno. Vico.. Joyce." *Disjecta: Miscellaneous Writings and a Dramatic Fragment.* London: John Calder, 1985.

Beckett, Samuel. The Collected Works of Samuel Beckett. New York: Grove, 1970.

Bell, Ian F. A., *Critic as Scientist.* London: Routledge, 1981.

Benjamin, Walter. *Illuminations.* Schocken: New York, 1968.

Bruno, Giordano. *On the Infinite Universe and Worlds.* New York: Henry Schumann, 1950.

Eliot, T. S. "Tradition and the Individual Talent." In *Selected Essays,* 3–11. New York: Harcourt, Brace, 1950.

Ellmann, Richard. *The Letters of James Joyce.* New York: Viking, 1966.

Fenollosa, Ernest, and Ezra Pound, *The Chinese Written Character as a Medium for Poetry: A Critical Edition,* Edited by Haun Saussy, Jonathan Stalling, and Lucas Klein (New York: Fordham University Press, 2008).

Flusser, Vilém. *Into the Universe of Technical Images.* Translated by Nancy Ann Roth. Minneapolis: University of Minnesota Press, 2011.

---. *Kommunikologie weiter denken: Die Bochumer Vorlesungen.* Frankfurt: Fischer, 2010.

---. *Towards a Philosophy of Photography.* London: Reaction, 2000.

---. *Writings.* Minneapolis, MN: University of Minnesota Press, 2002.

Goethe, Johann Wolfgang von. *Selected Poems.* New York: Penguin, 2005.

Helsinger, Howard. "Joyce and Dante." *ELH* 35, no. 4 (December 1968): 593.

Irrgang, Daniel. "Vilém Flusser in Open Circuits." *Understanding Flusser, Understanding Modernism.* New York: Bloomsbury, 2022.

Jaffe, Aaron. "The Double Repatriations of Cybernetics and Media Theory: Vilém Flusser and Heinz von Foerster." *New Literary History* 44, no. 2 (Spring 2023).

[69] Pound, 79.

Joyce, James. *Finnegans Wake*. New York: Viking, 1966.

---. *Portrait of the Artist as a Young Man*. New York, Huebsch, 1916.

---. "The Day of the Rabblement," *The Critical Writings of James Joyce*, ed. Ellsworth Mason & Richard Ellmann (NY: Viking Press 1959), 68.

Kenner, Hugh, *The Pound Era* (Berkeley, CA: University of California P, 1971), 160, 275-276.

---. *The Pound Era*. Berkeley, CA: University of California P, 1971.

Kindley, Evan. "Kenne's Networks," *Critical Inquiry* 47 (3):525-543 (2021).

Kittler, Friedrich, *Gramophone, Film, Typewriter* (Palo Alto, CA: Stanford, 1999), 46.

---. "The History of Communication Media," *Ctheory* (1996), https://journals.uvic.ca/index.php/ctheory/article/view/14325/5101.

Koyré, Alexandre. *From the Closed World to the Infinite Universe*. Baltimore, MD: Johns Hopkins University Press, 1957.

Martínez, Alberto A. "Was Giordano Bruno Burned at the Stake for Believing in Exoplanets?" *Scientific American*, March 19, 2018, https://blogs.scientificamerican.com/observations/was-giordano-bruno-burned-at-the-stake-for-believing-in-exoplanets.

Morton, Seth. *Modernist Interference: Inhuman Frequencies Between Modernism and Media*. Ph.D. diss. Rice University, 2019.

Murphet, Julian. *Multimedia Modernism*. Cambridge: Cambridge UP, 2009.

Pound, Ezra. "Henry James," *The Literary Essays of Ezra Pound*. London: Faber, 1954.

---. *Pound/Joyce: the Letters of Ezra Pound to James Joyce*. New York: New Directions, 1970.

---. *Selected Prose, 1909-1965*. New York: New Directions, 1975.

---. *The Cantos of Ezra Pound*. London: Faber and Faber, 1975.

---. *The Chinese Written Character as a Medium for Poetry: A Critical Edition*. Edited by Haun Saussy, Jonathan Stalling, and Lucas Klein. New York: Fordham University Press, 2008.

---. *The Literary Essays of Ezra Pound*. New York: New Directions, 1968.

Raschke, Daniel. "Flusser and Ars Electronica: Between and Beyond Cybernetics," *Understanding Flusser, Understanding Modernism*. New York: Bloomsbury, 2022.

---. Flusser on the Fly: Towards an English Translation of Flusser's Bochum Lectures," *Flusser Studies* 33, https://www.flusserstudies.net/sites/www.flusserstudies.net/files/media/attachments/raschke-flusser-on-the-fly.pdf.

Saussy, Haun. "Fenollosa Compounded: A Discrimination," *The Chinese Written Character as a Medium for Poetry: A Critical Edition*. New York: Fordham University Press, 2008.

Stein, William Bysshe. "Beckett's 'Whoroscope': Turdy Ooscopy," *ELH* 42, no. 1 (Spring 1975): 125-155.

Stevenson, Guy. "Fascism, Iconoclasm, and the Global Village." *Understanding Flusser, Understanding Modernism*. New York: Bloomsbury, 2022.

Trotter, David. *Paranoid Modernism*. Oxford University Press, 2001.

Verene, Donald Phillip. *James Joyce and the Philosophers at Finnegans Wake.* Evanston, IL: Northwestern University Press, 2016.

---. *Vico and Joyce.* Albany, NY: State University of New York Press, 1987.

Walkowitz, Rebecca. *Born Translated: The Contemporary Novel in an Age of World Literature.* New York: Columbia UP, 2015.

Washington University Image Collections and Exhibitions, http://omeka.wustl.edu/omeka/items/show/14843.

Woolf, Virginia, *The Diary of Virginia Woolf,* Volume II (New York: Harcourt, Brace, Jovanovich, 1980), 189.

Zielinski, Siegfried, *After the Media* (Minneapolis, MN: U of Minnesota.

Chapter 7
Scratching the Surface: Irma Vep's Analog and Digital Histories

Paula J. Massood
Brooklyn College, City University of New York

Abstract

From her earliest appearance in Louis Feuillade's *Les Vampires* (1915–1916) to her latest iteration in HBO's *Irma Vep* (2022), Irma has functioned not only as a symbol of criminal and sexual transgression but also as a springboard for examinations of filmic and digital materiality. Beginning in an analog and highly combustible nitrate form, Irma metamorphized over more than a century's worth of technological shifts and industrial globalization to become a digital and transnational container of the past and present, of history, of representation, and of the very ontology of cinema. In Olivier Assayas's *Irma Vep* (1996), for example, the role most famously played by French actress Musidora is performed by Hong Kong actress Maggie Cheung, the casting choice thus openly questioning definitions of national cinema. Likewise, the most recent streaming serialization further expands questions of nationalism and materiality by looking both back towards the historical roots of the character and French cinema and forwards, situating the narrative in a digital and global streaming present dominated by the Marvel Cinematic Universe and starring a Swedish actress. This essay argues that such shifts, makeovers, reboots, and remakes provide fertile ground for understanding cinema's shift from the nitrate film strip to the digital stream. I focus on three related moments featuring Irma Vep—from 1915, 1996, and 2022 respectively—to illustrate early film's persistent presence in a digital world and to argue that the analog is no more "real" than its digital counterpart.

Keywords: Film History, Analog Film, Digital Film, Remakes, Serials and Series, Global Franchises.

***

> There is no shortage of theoreticians who have attempted to ground the art of moving images on the solid base of the means specific to it. But the means specific to yesterday's analogical machine and to today's digital machine have shown themselves equally suitable for filming both

> love stories and abstract dances and forms. … *Cinema,* like *painting* and *literature,* is not just the name of an art whose processes can be deduced from the specificity of its material and technical apparatuses. … [C]inema is the name of an art whose meaning cuts across the borders between the arts.
>
> Jacques Rancière, *Film Fables* (2006).

From its beginnings, film theorists, historians, and filmmakers have repeatedly attempted to define cinema: its technological base, its ontological specificity, its status as a language (or not), its psychological effects and ideologies, or its death (and rise again) in the digital age. Such examinations have been complicated, at least during film's early years, by two factors, cinema's photographic base and its aesthetic relationships to vaudeville and low-brow amusements, grounding it in both realism and fantasy, respectively. For theorists like Siegfried Kracauer and André Bazin, cinema's defining characteristics were enabled first and foremost by its photographic foundations, which linked the emerging art to material reality: there was someone or something in front of a camera. Kracauer argued in his *Theory of Film,* for example, that film's basic properties were identical to the properties of photography. As he suggests, "film, in other words, is uniquely equipped to record and reveal physical reality and, hence, gravitates toward it."[1] Indeed, "the nature of photography survives in that of film,"[2] and enables its ability to not only record material reality, but to reveal components of that reality to audiences who may overlook it in the world. For Kracauer, film was the logical end to the "age old desire to picture things moving."[3] It was an expansion of the image-world already being experienced with photography, providing "a haunting, posthuman embodiment of not presence but absence, not life but death; or perhaps more accurately, life in the face of death."[4]

While Kracauer is often considered to be one of the major early scholars of film realism, theorizing as he did from the 1920s onward, André Bazin, writing in the aftermath of World War II, also focused on cinema's debt to its artistic forebears, including painting and photography. In "The Ontology of the Photographic Image" (1945), for example, Bazin famously (and contentiously) argued that the arrival of photographically based film was "the first time an

---

[1] Siegfried Kracauer, *Theory of Film: The Redemption of Physical Reality.* (New York: Oxford University Press, 1960), p. 28.

[2] Ibid, p. 27.

[3] Ibid.

[4] Paula Amad, "Film as the 'Skin of History': André Bazin and the Specter of the Archive and Death in Nichole Vedres's *Paris 1900* (1947)," *Representations* 130 (Spring, 2015), p. 100.

image of the world is formed automatically, without the creative interventions of man ... All the arts are based on the presence of man, only photography derives an advantage from his absence."[5] For Bazin, cinema's mechanical (e.g., photographic) base contributed to film's artistic elements by providing audiences with the freedom to look and explore the world contained within the image, which accounts for his famous predilection toward deep focus and long takes over montage.

In the past, Bazin's positions were often critiqued as naïve realism, and yet his understanding of the links between material reality and film, via photography and cinematic aesthetics—such as editing, cinematography, and mise-en-scene—offer a far more complex approach to our ongoing negotiations between material reality and representations of reality, for what is cinema if not a representation of an a priori materiality in both its analog and digital forms? At the same time, what does it mean when the image seen—whether analog or digital—is present but its referent exists as an absent presence? As Paula Amad and others suggest, this absent/presence in Bazin's writing can be connected to the linkages he made between film and history wherein, "Bazin emerges as a theorist of the image in an archival age, and rather than merely a historian of film style, he becomes a theorist of film style *as* history."[6] As the following will argue, Bazin's understanding of film and history pertains to the character of Irma Vep and her multiple reincarnations.

More recent scholarship on Bazin, like Amad's above, has opened up his work in light of contemporary philosophy, film theory, and new technologies. Bazin was a critic first and foremost and many of his essays suggest a flexibility when defining cinematic realism; for example, he was as accepting of documentary and experimental film as he was to classical Hollywood and Italian Neorealism because, as he argued, "the screen reflects the ebb and flow of our imagination, which feeds on a reality for which it plans to substitute. ... Correspondingly, what is imaginary on the screen must have the spatial density of something real."[7] Realism, for Bazin, assumes a material reality and just may be cinema's greatest fiction. Likewise, until his early death in 1958, Bazin was a curious explorer and sometime exponent of "new technologies" (his term), including Cinerama, Cinemascope, 3D, and television, and he wrote extensively on these developments during the 1950s. In recently reconsidering Bazin's later writings,

---

[5] André Bazin, "The Ontology of the Photographic Image," *What is Cinema? Volume 1.* (Berkeley, CA: University of California Press, 2014), p. 13.

[6] Amad, p. 88.

[7] André Bazin, "The Virtues and Limitations of Montage," *What is Cinema? Volume 1.* (Berkeley, CA: University of California Press, 2014), p. 48.

particularly those focused on television, Dudley Andrew argues that the critic believed that cinema's "distinct ontology … derives from its being an image that, eerily, is both from the past and alive: life deferred indefinitely, re-animated at every projection." Andrew argues that Bazin's "image family" included photographs, films, and television, with the latter, at the time he was writing, "depending on the liveness of present perception." [8] Thus, new media created, for Bazin, the ability to reconsider the relationship between the image and time, materiality and the screen.

What does mid-century realist film theory have to do with the character of Irma Vep from Louis Feuillades' *Les Vampires*, a silent French film serial from cinema's beginnings? Furthermore, what does she, and the serial, have to do with her late-20th and early-21st-century digital, and in the latter case streaming, reincarnations? As Jacques Ranciére suggests in the opening epigraph, film's perceptual affects extend beyond "the solid base of the means specific to it." Indeed, Ranciére is arguing for a fluidity between technological base and image, where the latter becomes "aesthetic acts" that create "new modes of sense perception." Irma Vep—a very French silent film antihero—provides an apt example of a figure who began as an analog shadow etched into a nitrate negative but has morphed over time and technologies; from analog serial projected in music halls and other places of mass entertainment, to digital feature film projected digitally in multiplexes, to television serial, streamed on televisions, computers, and phones.

***

> The dilemma posed by the figure of Musidora [as Irma Vep] circulates in the domain of the metacinematic and metaphysical: what do we know from the image (and how does that parallel or exceed human vision), and, more fundamentally, what can we know at all?[9]

In the penultimate episode of the HBO series, *Irma Vep* (2022), actress Mira Harberg visits director René Vidal to request that he return to the set of the television show they've been shooting in Paris: already mentally fragile, he's suffered a stress-related breakdown elicited by the demanding shoot. The "series within the series" is a remake of the ten-part silent film serial, *Les Vampires,* directed by Louis Feuillade for Gaumont in 1915–1916. Mira, an American actress best known for her roles in Marvel action films, is cast against type as Irma Vep, a femme fatale role originally played by the French actress

---

[8] Dudley Andrew, "Introduction: Andre Bazin Meets the New Media of the 1950s," *Andre Bazin's New Media.* (Oakland, CA: University of CA Press, 2014), p. 8.

[9] Vicki Callahan, "Screening Musidora: Inscribing Indeterminacy in Film History," *Camera Obscura* 48 (16.3, 2001), p. 70.

and vamp, Musidora. She isn't the second Irma Vep; indeed, she's the third, following another "Irma" played by Maggie Cheung in Olivier Assayas's 1996 feature film of the same name. The HBO version, also directed by Assayas, is just one of Irma's many afterlives, with her latest portrayal literally haunted by these earlier renderings.

From her earliest appearance in Feuillade's *Les Vampires* to her latest iteration in the HBO show, Irma has functioned not only as a symbol of criminal and sexual transgression—a shape-shifting fugitive from the law—but also as a springboard for examinations of filmic and digital materiality, as visualized by Assayas. Beginning in an analog and highly combustible nitrate form—the form more dangerous than the character[10]—Irma metamorphized over more than a century's worth of technological shifts and industrial globalization to become a digital and transnational container of the past and present, of history, of representation, and of the very ontology of cinema. In the 1996 film, for example, the role most famously played by Musidora is performed by Hong Kong actress Maggie Cheung, Assayas's casting choice thus openly questioning definitions of national cinema while also—in an ending sequence shot in 35mm—referencing film's material base, a point to which I'll return.[11] Likewise, the most recent streaming serialization further expands questions of nationalism and materiality by looking both backwards towards the historical roots of the character in particular and French cinema more generally and forwards, situating the narrative in a digital and global streaming present dominated by the Marvel Cinematic Universe and starring a Swedish actress most well-known for her work in American film. This essay argues that such shifts, makeovers, reboots, and remakes provide fertile ground for understanding cinema's shift from the nitrate film strip to the digital stream. In what follows, I focus on three related moments featuring Irma Vep—from 1915, 1996, and

---

[10] For more on early attitudes towards the handling of early film stock, see Herve Picherit, "Making a Modernist Masterpiece: The Elusive Location of Cultural Significance in *Les Vampires*," *Modernist/modernity* 25.4 (November 2018), pp. 655–681. According to Picherit, referencing scholar Mathilde Mansoz, "film stock, including negatives, was regulated as industrial waste. It was illegal to stock more than a certain quantity of film in one place, such that the studies dispersed their negatives in different storage sites ... many film companies resorted to a more definitive solution, burning or dumping their film stock into rivers and lakes ... the very act of keeping film negatives, then, was often the unintended outcome of the need to store dangerously flammable film base off-site of the studio" (659). Such practices often resulted in lost or fragmented negatives, including those for *Les Vampires*, which was first reassembled and restored in 1984.

[11] Coincidentally, 1996 was the first year an all-digital film was released, and Netflix was founded, thus beginning of the age of digital streaming services.

2022, respectively—to illustrate early film's persistent presence in a digital world and to argue that the analog is no more "real" than its digital counterpart. Do filmic images, as Vilém Flusser argues "absorb the whole of history and form a collective memory going endlessly round in circles"?[12] Alternately, do they suggest more of Roland Barthes argument regarding cinema's "protensity"; its continuity and flow, the latter an apt term as well for serialized television.[12] Finally, do they prove the "death of cinema," as many film historians have feared, or do they suggest an expansion of the visual predicted by Bazin in his writing on new technologies and television?

First, a bit of film history: *Les Vampires* was the second installment in a trilogy of crime serials made by Feuillade for Gaumont during World War 1, along with *Fantomas* (1913) and *Judex* (1916). Crime novels and printed and filmed serializations were popular at this time in France and elsewhere and were influenced by "conservative suspicions of urban sites, especially around fears of the fallen or "fatal" woman.[13] In other words, they were transmedial narratives that appealed to audiences who were experiencing the dislocations and shocks of modernity, which included rapid industrialization and the growth of urban areas. Concurrently, Paris was experiencing a crime wave, and such narratives had the effect of linking criminality and burgeoning urban spaces in early film history.[14]

Les Vampires focuses on a group of anarchic underworld criminals, or "Apaches,"[15] who, dressed in hooded black outfits, terrorize Paris's bourgeoisie through murder, burglary, and kidnapping. The gang is led by a male "Grand Vampire" along with Irma Vep, to whom he is often romantically linked. The serial was shot on the streets and rooftops of an eerily empty bomb-struck Paris, on black and white film stock that was often toned to emphasize time

---

[12] Roland Barthes, *Camera Lucida*. Trans. Richard Howard. (London: Flamingo, 1984), p. 90.

[13] Kristen J. Butler, "Irma Vep, Vamp in the City: Mapping the Criminal Feminine in Early French Serials," in *A Feminist Reader in Early Cinema*. Eds. Jennifer M. Bean and Diane Negra. (Durham, NC: Duke University Press, 2002), p. 210.

[14] Paris was not the only city to experience the dual effects of industrialization and criminality (perceived or real). New York City films, for example linked the two elements very early, with conventions of the gangster genre appearing onscreen even earlier than the serials. See, for example, *The Black Hand* (Biograph, 1906). One difference, however, is that the American films xenophobically linked criminality to the nation's burgeoning immigrant population, Italians, in the case of the Biograph film.

[15] The term apache refers to Parisian street gangs and underworld criminals. While its origins are unclear, it is thought to refer to Native American tribe, telescoping racist fears of savagery onto France's criminal and working classes. In 1904 Eugene Grecourt described the term as referring to "gangs of pimps and criminals" in *L'Intermediaire des Chercheurs et Curieux* 1032. 279 (29 Feb. 1904), p. NA.

(day/night) and content (for example, danger). Because male actors were regularly called away to fight in the War, the performers playing the Grand Vampire often changed, leaving Irma Vep as the most consistent element in the series. The result is that she is most often regarded as its star, an impression supported by the serial's promotional materials (lobby cards, posters, and ads) which prominently featured her black-clad figure. The gang is pursued by a journalist, who works with the police to identify and arrest its fugitive members, particularly Irma, and the narrative clearly emphasizes the erotic, transgressive, and criminal aspects of the story wherein Irma is a bad woman who needs to be punished.

Musidora, a former dancehall performer (writer and future film director), gave Irma her agency, allure, and unabashed sensuality; in all creating an embodied glorification of gendered and urban criminality. Moreover, her physicality and daring exploits provided for a reimagining of the figure of the flañeuse because her movements around the city streets and rooftops are free of the dangers and social constraints that limited bourgeois women of the time. In fact, her criminal motives (in this case, non-sexual as they are focused on theft) make Irma more of a threat. Musidora had appeared in films prior to her work with Feuillade, but it was her role in *Les Vampires* that catapulted her into stardom, a rare phenomenon for French (and American) cinema at the time, when companies resisted crediting performers out of fear of rising salaries. Feuillade, coming off the success of *Fantomas,*[16] began work on *Les Vampires,* which was Gaumont's answer to the popularity of American serials, such as *The Perils of Pauline* (1914), a popular action series produced by Paramount Pictures, which also featured a brave heroine. *Les Vampires* was intended to compete with Pathe's[17] *Les Mystères de New York* (1914–15), scheduled for release in France in 1915. The French serials were produced at a time when the industry was suffering from the economic and social effects of the war and increased competition from American films, which filled the gap on French screens. Gaumont's releases were intended to help the industry and company survive. Moreover, argues Dale Hudson, "[b]oth Irma Vep and Musidora … functioned as national icons … that inspired the admiration of French surrealists, such as

---

[16] *Fantomas* was an adaptation of a series of crime novels by Marcel Allain and Pierre Souvestre. The eponymous lead is a masked criminal mastermind, much like the Grand Vampire and Irma Vep in *Les Vampires.*

[17] Pathe was the American division of the French film company, Pathe Freres. Pathe was established in 1914 and had studios in New Jersey.

Louis Aragon,"[18] who wrote about the film in retrospect and was drawn to the serial's ephemeral nature (by the time they were writing, the prints had all but disappeared).[19] In a related sense, Surrealism played a part in Bazin's theories of realism a few decades later. The movement, and especially its use of photography, "eliminated ... the logical distinction between the imaginary and the real," wherein "every image should be experienced as an object and every object as an image."[20]

Irma Vep, defined by Musidora's sensuous performance, has been the subject of various responses since she first appeared on screen, with particular emphasis placed on her identity-concealing, yet body-revealing black silk body suit (an object as well as an image), which was much more revealing than that worn by her male counterparts. In Episode 6, "Hypnotic Eyes" (1916), for example, we see Irma don a black mask and catsuit and sneak through the halls of a hotel before entering a victim's room to steal a treasure map. She's caught by a rival criminal, Moréno, who unmasks her before hypnotizing and kidnaping her and replacing her with another woman (whose suit is far less revealing). Kristen Butler argues that Irma Vep is "best remembered in her black body suit [despite it making relatively few appearances in the film] ... an inky character fetish that simultaneously highlights the visible evidence of her gender and poses the problem of visibility" by blending into the shadows of the black and white film stock and reminding us of the fungibility of the image.[21] Musidora designed the outfit of almost translucent silk, a look and fit that resulted in a dark silhouette that Butler describes as simultaneously a "*tache* or stain on the screen" and a "glyph" or marker that serves many [sexual and criminal] purposes."[22] Through costuming and behavior, therefore, Irma serves as a "free-floating sexual signifier"[23] that crosses boundaries, literally and figuratively, by entering forbidden spaces, changing identities, and even cross-dressing (as when she wears another costume to pass as a young Viscount to the Grand Vampire's Count disguise earlier in the episode). Her dual status as liminal and yet omnipresent—as

---

[18] Dale Hudson, 'Just Play Yourself, Maggie Cheung': *Irma Vep,* Rethinking Transnational Stardrom and Unthinking National Cinemas," *Screen* 47.2 (Summer 2006), p. 219.

[19] Herve Picherit, "Making a Modernist Masterpiece: The elusive Location of Cultural Significance in *Les Vampires.*" *Modernism/modernity* 25.4 (November 2018), p. 663.

[20] Bazin, pp. 9–10.

[21] Butler, p. 201. Musidora appears in costume earlier in the series, but only as one among many Apaches, the form fitting suit identifying her from the te crowd of loose fitting and androgenous disguises of the others.

[22] Ibid, p. 206.

[23] Ibid, p. 204.

nowhere and everywhere, as no one and everyone[24]—returns us to the question of the status of the analog and the digital, for what happens when an object makes a brief appearance in front of a camera and then takes on a virtual like of its own?

Fig. 7.1: Musidora in disguise, searching for the map in Moréno's hotel room.

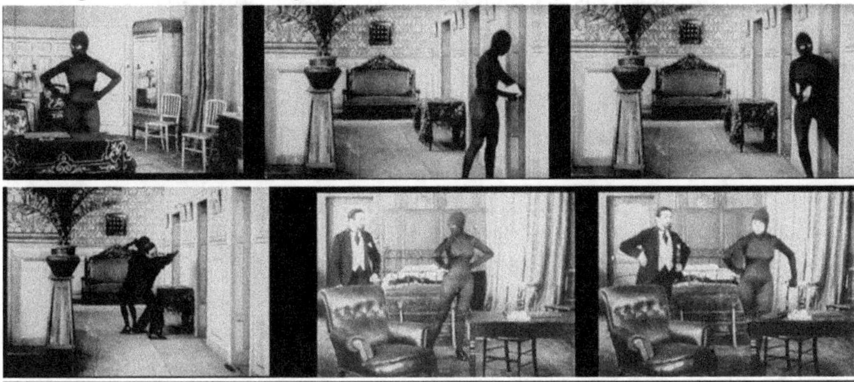

In her study of Musidora, film historian Vicki Callahan argues that Irma's "black silk costume circulates around almost every discussion and later filmic invocation[s] of the figure," becoming "an embodied and mobile meta-cinematic devise floating through the films."[25] While Callahan is referring specifically to *Les Vampires*, I'd like to pursue this line of argument through two more iterations of the character by focusing on the role she plays in the feature film and its streaming remake. I suggest that Irma Vep functions almost like a chronotope, telegraphing film history through a particularly gendered form that relies on the iconicity of the original, while radiating outwards from there. Neither the film nor the television series is a true remake of Feuillade's original, nor are they crime films per se—though each contain scenes of the actresses committing crimes while dressed as Irma. Instead, Assayas's film is a movie about a movie, a comedic meta-film about the attempt by a troubled director named René Vidal (Jean-Pierre Leaud) to remake Feiullarde's original *Les Vampires* in an updated serial form for French television. The second *Irma Vep* follows in the first one's footsteps by narrating the attempts of a French filmmaker, also named René Vidal (Vincent Macaigne), to remake *Les Vampires*

[24] The series is also known for another black-clad woman who appears in the first and second episodes. This figure is Marfa Koutilof (Stacia Napierkowska), a dancer who wears a bat costume and descends onto the stage during a performance. She dies on stage, having been poisoned by the Grand Vampire. Her murder is the inspiration for the journalist's investigation.

[25] Callahan, p. 73.

as streaming television intended for a global viewership. It is also a loose remake of the 1996 film and incorporates clips from that source as well as the original serial. It also includes scenes set in 1915–1916, adding a historical element to the contemporary framing narrative. In both versions, the cast and crew are multi-national and multi-lingual, emphasizing the changes in French film—and filmmaking more generally—in the late 20th and early 21st centuries, reflecting a globalization that has blurred the boundaries between national film and media industries.

Irma Vep's director, Olivier Assayas, is a direct descendant of the French New Wave. A critic for *Cahiers du Cinema*—like its founding editor André Bazin—he has written extensively on European film. In addition, he has authored book-length studies of Asian cinema, and these aesthetic and historical influences can be traced throughout both *Irma Veps,* accentuating the works already pronounced *cinephilia,* especially its love of the image; its surfaces and visual possibilities. Rather than focusing on updating the original narrative, Assayas employs the character of Irma as a starting point for an examination of filmmaking—past and present, national and transnational, analog and digital. René, as Dale Hudson suggests, "is remaking a cinematic 'archetype,' the character of Irma Vep, not *Les Vampires,* so that *Irma Vep* becomes a film about how [an archetype] has to be remade for every generation."[26] This focus enabled Assayas to avoid expectations of fidelity to Feuillade's original and provided an opportunity to explore the ontology of the image in a digital context, analyzing the status of the cinema through the character. As part of this project, Assayas emphasizes film's analog roots by intercutting Super 16mm with 35mm with film stock. For the director, such aesthetic and narrative bricolage celebrates the cinema's fluidity rather than mourning its erasure in the face of new technologies; and what is cinema if not always already a new technology?

Annette Forster observes that the criticism around *Les Vampires* (its criminality, for example) elicited a "concern about the future of French cinema,"[27] and its devolution into popular entertainments drawn from newspaper serials and music halls. While such critiques refer to the serial's criminal and amoral content, there were also fears about the economic future of the industry, fueled by France's wartime financial struggles and American cinema's increased presence in French theaters. Likewise, Assayas's *Irma Vep,* self-consciously elicits similar questions about the state of French cinema; first through the character of René, whose casting of Jean-Pierre Leaud, primarily known for his

[26] Hudson, p. 29.

[27] Annette Förster, *Women in the Silent Cinema: Histories of Fame and Fate.* (Amsterdam: Amsterdam University Press, 2017), p. 227.

work with Francois Truffaut and Jean-Luc Godard, and filmmaking style references the Nouvelle Vague, a recognized apex in cinema history. Second, and more radically in the choice of Maggie Cheung, a Hong Kong action film star, for the role of Maggie, the actress playing Irma Vep—one that made extradiegetic sense given Assayas's interests.[28] Cheung's casting was controversial within and outside of the film, especially given Musidora's and Irma's standing in the national mythos. Irma is French, and Maggie clearly is not, and the narrative plays on her disruptive status as a transnational film star of Chinese descent, with Assayas clearly critiquing the notion of cinematic purity.

Much of the film is taken up by characters—cast, crew, reporters—discussing the nature of filmmaking, from French film to Hollywood and Hong Kong, and in this way, the film appeals to an audience of cinephiles. It is in these discussions where the porosity of cinema's aesthetics, technologies, and personnel are highlighted. Just as much time is devoted to Irma Vep, who haunts the film's explorations of cinema and whose history is signified by the black suit worn by Maggie in the film (and the film within the film). Unlike Musidora's translucent silk, Maggie's costume is black latex and, in one scene, we see her fitted for the suit in a BDSM shop. In this example, Assayas draws humor from his casting choice by placing her entirely out of context while also pointedly referencing the underground sensuality of Irma's past and present costumes. Maggie wears the outfit through much of the film, even when she isn't on set, totaling far more screen time than Musidora's original (*Les Vampires* was over seven hours long while *Irma Vep* is 96 minutes). We watch while she creeps around her hotel in disguise, voyeuristically spying on guests and stealing jewelry out of a room, not only mimicking Musidora's/Irma's actions but also mirroring an earlier scene in which René films Maggie/Irma recreating the hotel scene from *Les Vampires.* It also suggests that Irma Vep's ghost (via the suit) encourages rebellious behavior: Maggie is far lither and more sensual while in the suit than when in her street clothes, when she appears clumsy and awkward. This is noteworthy given the fact that René hired her based on her choreographed feats in her Hong Kong films, which Maggie explains were performed by a body double. Given that Maggie's face is clearly exposed in her *Irma Vep* scenes, we understand that the suit acts as a body double of sorts, providing her with a physicality much more in keeping with her Hong Kong persona.

In Feuillade's original, Irma Vep's movements remained silent, with either live or recorded music providing a real or analog (on sound discs, for example) accompaniment to the narrative. This silence provides Irma Vep with a menacing visual presence—a *tache* or a *glyph*—because the audience is not being led by

---

[28] As Maggie, Cheung is playing a fictional version of herself. For more on this see Dale Hudson.

sound; the suturing into the narrative is solely visual. Spectators are asked to watch the character creep through hallways and run across Paris's rooftops with the actual city in clear sight in the background, giving Irma a superhuman, uncanny affect. In Assayas's version, on the other hand, the latex suit squeaks as a part of the diegetic sound, forcing the crew to try to eliminate its unruly disruption. In scenes of the crew shooting the film, this element is comedic and distracting, functioning as Assayas's gesture towards Feuillade's silent version, while also serving as a reminder of the visuals' powerful effects. In Rene's rough cut, revealed later in the film, the sounds are removed, to be replaced with discordant sound effects reference film's visual and sonic surfaces.

Les Vampires is never remade; René suffers a breakdown and cannot finish the project. He's replaced by another filmmaker, Jose Murano, who immediately recasts Irma with a French actress because, as he says, Irma is quintessentially French and should not be played by "La Chinoise," thereby negating Rene's casting choice.[29] And yet, we glimpse Rene's intended version in his unfinished cut. Near the end of *Irma Vep*, the cast and crew watch René's cut. The two-minute mini-film neither resembles *Les Vampires* nor the content of the few scenes shown being filmed in *Irma Vep*. Rather than a conventional narrative, René's version of *Les Vampires* is a rapid montage of black and white footage of Maggie dressed as Irma, climbing across the rooftops of Paris, wandering hotel halls, and being hypnotized by Moréno. The film's surfaces are etched with nonsensical doodles, letters, and shapes, and are either superimposed over each other or intercut with white screens. The soundtrack alternates between the diegetic sounds of the group in the screening room and the cut's metadiegetic (sound emanating from the film within the film) bursts of electronic noise, acting as a shocking accompaniment to the scratches and punctuating the violence to the surface of the image. While some of the image degradation may remind audiences of the fragility of *Les Vampires'* film stock, the cut recalls the experimental films of French writer/filmmaker Jean Isidore Isou and the mid-century Lettrist movement, whose intention was to "destroy cinema's existing rules, with experimental films that combined scratched images with discordant noise. [30] While Rene's film may not have as radical an intention as what Isou produced, the manipulated and impure footage disrupts expectations of a

[29] This casting is an intentional irony on Assayas's part given that the replacement is played by Lou Castel, a Columbian born, Swedish actor who appeared in films by Luchino Visconti, Rainer Warner Fassbinder, Wim Wenders, Claude Chabrol, and others. His casting is especially resonant given that in Fassbinder's *Beware of a Holy Whore* (1971), he played an unstable and disruptive director of a film that is never finished.

[30] <<https://www.spectacletheater.com/lettrist-cinema/>> Accessed 11 Jan., 2924. American experiment filmmaker, Stan Brakhage, also took inspiration from Isou.

conventional remake. Moreover, it calls attention to the "skin of the film," enabling a sort of haptic visuality that "moves over the surface of its object rather than [plunging] into illusionistic depth, not to distinguish form so much as to discern texture;" a characteristic of film assumed lost with digital technologies, which are often pristinely clean.[31]

Fig 7.2: Rene's final cut 1996, revealing the surface of the image.

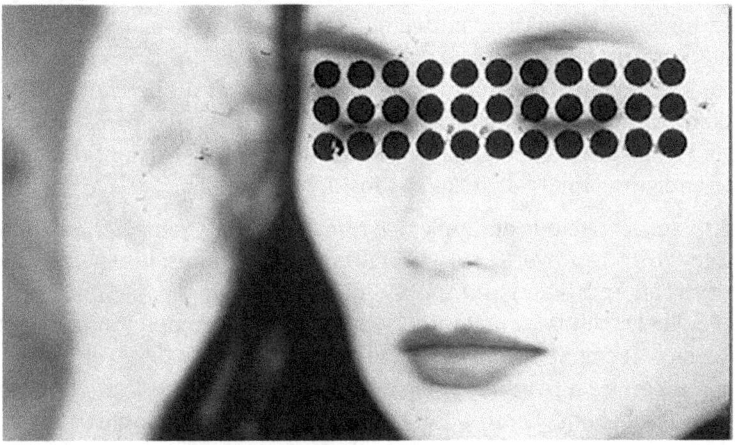

Assayas's streaming series, *Irma Vep*, is a remake and extension of his earlier film. Its serial format—eight episodes—provided the director with the time and space to explore French film's past through more expansive references to *Les Vampires* and his previous *Irma Vep*, while also covering familiar narrative ground involving the character of Irma Vep and a director's attempts at a remake. In many ways, the series follows the themes and narratives of the original film, with the emphasis on the character of Irma and the actress cast in the role, here renamed Mira Harberg and played by Alicia Vikander. René appears again as a troubled director trying to remake Feuillade's serial while also being haunted by his experiences (and Maggie) with the earlier film. He suffers a breakdown and leaves the film, to be replaced by an American director of Marvel action

[31] Laura U. Marks, *The Skin of the Film: Intercultural Cinema, Embodiment, and the Senses*. (Durham, NC: Duke University Press, 162. Early film stock wore its age and usage on the surface of the print. One of the ironies of archival restorations is that the process often digitally erases a history present in a print's scratches, watermarks, tears, and splices. Moreover, most restorations result in digital copies, with the rare celluloid prints struck from digital negatives. Film restorations such as that performed on *Les Vampires* result in beautiful prints that are far removed their original nitrate base. Ironically, attempts to bring audiences closer to the original provide a high-quality digital image of what we believe that original to be.

films, whose main change is to insert CGI special effects and drone shots into a film that is essentially a period piece.[32] The cast and crew—expansively multi-lingual and multi-national—continue to puzzle over the contemporary state of global filmmaking, exuding an "anxiety" about "a cinema world that has changed much since the time of [Assayas's] original movie" and *Les Vampires* before that.[33] Indeed, the series suggests that to be a filmmaker in 2022 or 2023 one must work in digitally streaming serial narratives—ironically returning to a version of the analog past in the digital present—and understand the importance of cross-platform branding. And yet, as one character, Gottfried (Lars Eidinger), reminds the cast and crew, and therefore the show's audience: "The industry has taken over cinema. Lawyers, big data, franchises, platforms, you name it. But indie films are no better. They preach until you are sick of them. Cinema was for bad guys, and bad girls, like Musidora."

With its self-conscious and reflexive remake about a remake about a remake structure, *Irma Vep* overflows with cinematic self-consciousness, with Irma serving as "an embodied and mobile metacinematic device floating through these [Feuillade's] films," along with, I would argue, their contemporary iterations. Assayas also recognized the character's mutability: "whenever cinema is in chaos, you can do a new *Irma Vep*, and I felt pretty much that cinema was in the moment of chaos. There was a lot of soul searching, people trying to figure out what's going on [during COVID]."[34] But, as Gottfried's speech suggests, cinema was in a moment of change that extended beyond the limitations imposed by a global pandemic.

Like Musidora and Maggie, Mira wears a version of Irma's iconic costume, this one made from a plush velvet fabric. She reenacts many of the scenes from *Les Vampires*—more than Maggie—and *Irma Vep*, creeping across Parisian rooftops, sneaking through hotel hallways and rooms, slipping into adjoining apartments, and shapeshifting through walls and ceilings; a flaneuse and jewelry thief much like her predecessors. Unlike its previous filmic iteration, however, this *Irma Vep* is haunted by the ghosts of several pasts. First, in flashback-like scenes detailing *Les Vampires* historical context, much of it drawn from Musidora's memoirs. Second, by the ghosts of Assayas's history with the 1996 film and his short-lived marriage to Maggie Cheung. Both Musidora's and Maggie's presence are complex, as they acknowledge film history as well as personal history, fiction

---

[32] While the film shoot functions as *Irma Vep*'s framing device, the film within a film is set in the early 20th and accords with the time frame of the original *Les Vampires*.

[33] Scott McCauley, "'Whenever Cinema is in Chaos, You Can Do a New *Irma Vep*": Olivier Assayas on His New HBO Series. *Filmmaker.com* (23 June, 2022). << https://filmmaker magazine.com/115163-olivier-assayas-irma-vep-hbo/>> (Accessed Oct. 2022).

[34] Ibid.

and nonfiction, with an affect, best described by Paula Amad in another context, as "a haunting, posthuman embodiment of not presence but absence, not life but death; or, perhaps more accurately, life in the face of death."[35]As Assayas stated: "Irma Vep, beyond Feuillade's movie, has a life of its own .... It's a ghost that has been handholding cinema ever since, and my movie is part of the story of Irma Vep."[36] It is also Assayas's story of the resilience of cinema in the face of a surmised digital death.

As a conclusion, let's return to *Irma Vep*'s penultimate episode, "The Horrible Marriage": Mira visits René, asking him to return to the series. She appears in his apartment, wearing the black velvet suit and floating down from the ceiling, having acquired the superhuman ability to pass through solid walls and float—though not in the superhero sense (for which Mira is normally cast). Her abilities to float and walk through walls have been initiated by René's earlier encouragement that she let herself be taken over by the ghost of the original Irma, thus adding elements of film's most basic attractions—magic and fantasy—to the show's mainly realist skeleton. When she wears the suit, Mira becomes Irma, allowing her to move through space effortlessly, a cinematic trick of sorts. This sense of fantasy, or hallucination, works differently here than in Visser's earlier pessimistic observation regarding the unanchored digital image because, in a sense, Mira *becomes* her analog counterpart (i.e., Irma), with Assayas updating Feuillade's low-tech stunts with CGI and other special effects.

Fig. 7.3: Rene's final cut 2022.

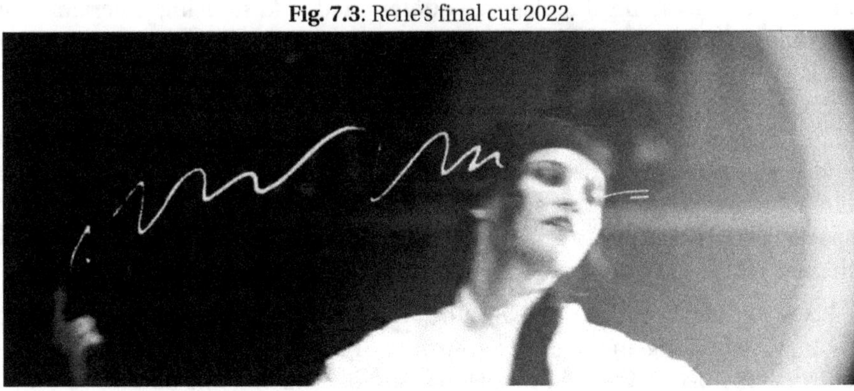

[35] Amad, p. 100. Here Amad builds from the writings of Siegfried Kracauer, Andre Bazin, and Virginia Woolf to understand the effect of archival footage in collage films which present "a glimpse of a world that does not include us and that now unfolds in a time that does not include many of the filmed subjects (100). A scholar of the archive, Amad is drawn to Kracauer and Bazin's reflections on memory, history and the nonfiction image.

[36] Assayas, qt. in McCauley.

During the latter part of the scene and before he becomes aware of Mira's presence, René screens an edit of the footage shot thus far. In some ways, this roughcut echoes that from the 1996 film; for example, it's a montage of rushes, edited together in a rapid rhythm and accompanied by discordant sounds and punk music. Images are scratched and superimposed, and the footage alternates between black and white, color, and saturated tones, including color-toned clips from the original series. It includes flickers and jumps of the frame, providing a spectatorial effect much like a film strip running through a projector (even though it's digitally projected). Clips from *Les Vampires* and the earlier *Irma Vep* are intercut into this version, resulting in a multimedia experience, combining film and digital video. This, along with the addition of counter heads and tails, the sound of a projector, and the size of the screen provides a multisensory and haptic experience that, to quote Marks, "forces the viewer to contemplate the image itself instead of being pulled into the narrative."[37] In this sense, Irma Vep shape shifts again, existing as both the depth and surface of the image through her analog and digital incarnations.

In 1947, and again in revised form in 1958, Bazin reviewed *Paris 1900* (Nicole Védrès, 1947), a compilation film constructed of unidentified found footage of mostly mundane events that, on the face of it, narrated a history of France's Belle Epoque period through the beginning of World War I (coincidentally abutting Les Vampire's historical moment). While Bazin was not an enthusiastic supporter of montage, he did not wholly discount it, and he saw in the film's juxtaposition of archival footage, to quote Amad, a "distancing experience, or defamiliarization" that he associated "with a tear (*déchirement*) in the film's surface or skin." Such tears, argued Bazin, create "an opening between 'chance and reality' ... one that was responsive to an appreciation of the cinematic image as simultaneously 'real' and 'fantastic,' factual and hallucinatory, present and past, always in flux." [38] Bazin's thoughts on the thickness and fragility of film's surface suggests that such self-conscious rendering of images can jolt spectators, forcing them to experience history and memory as something beyond themselves; as collective rather than personal.

Likewise, Rene's rough cuts, both in the 1996 and 2022 versions, juxtapose footage, scratching and tearing the surface of the film, resulting in a bricolage of Irma's history, of French film history, and of transnational and transmedial cinema more broadly. As he observes to Mira in this scene, Irma "is a spirit . . . cinema awakened her . . . and she [has] lingered ... haunting cinema for over a century." As this essay has argued, this haunting takes the form of multiple appearances and references, through which, Bazin argues elsewhere, "cinema

[37] Marks. p. 163.

[38] Amad, pp. 100–101.

has the exorbitant privilege of repeating [memory, history, an original moment], a privilege common to all mechanical arts [including photography], but one that it can use with infinitely greater potential than records or radio." To explicate, he continues, "I cannot repeat a single moment of my life, but cinema can repeat any one of those moments indefinity before my eyes."[39] Indeed, such cinematic repetition can act as a haunting, especially when it reveals collective memories. Contemporary cinema's detractors and cinephiles alike often bemoan the death of the cinema in the age of the digital. Assayas's versions of the character of Irma Vep suggest something less melancholic: they are not so much a nostalgic desire to recover a past—of French cinema, of celluloid—as they are a reminder of the resilience and power of images. As Ranciére argues, a longstanding lamentation in contemporary thought wants us to bear witness to the programmed death of images at the hands of the machine for information and advertisement. ... [but] the art and thought of images have always been nourished by all that thwarts them."[40] Irma Vep's repeated, transmedial transformations provide proof to Ranciére's pudding.

Bibliography

Amad, Paula. "Film as the 'Skin of History': André Bazin and the Specter of the Archive and Death in Nichole Vedres's Paris 1900 (1947)," *Representations* 130 (Spring, 2015), pp. 84-118. https://doi.org/10.1525/rep.2015.130.1.84.

Andrew, Dudley. "Introduction: Andre Bazin Meets the New Media of the 1950s," *Andre Bazin's New Media.* (Oakland, CA: University of CA Press, 2014), p. 8.

Barthes, Roland. *Camera Lucida.* Trans. Richard Howard. London: Flamingo, 1984.

Bazin, André. "Death Every Afternoon," in *Rites of Realism: Essays on Corporeal Cinema.* Ed. Ivone Margulies. Durham, NC: Duke University Press, 2002), p. 30. [27-31]. https://doi.org/10.2307/j.ctv123x744.5.

----. *What is Cinema?* Volume 1. Berkeley, CA: University of California Press, 2014.

Butler, Kristen J. "Irma Vep, Vamp in the City: Mapping the Criminal Feminine in Early French Serials," in *A Feminist Reader in Early Cinema.* Eds. Jennifer M. Bean and Diane Negra. Durham, NC: Duke University Press, 2002, pp. 195-220. https://doi.org/10.2307/j.ctv111jjcz.11.

Callahan, Vicki. "Screening Musidora: Inscribing Indeterminacy in Film History," *Camera Obscura* 48 (16.3, 2001), pp. 58-81. https://doi.org/10.1215/02705346-16-3_48-59.

[39] Andre Bazin, "Death Every Afternoon," in *Rites of Realism: Essays on Corporeal Cinema.* Ed. Ivone Margulies. Durham, NC: Duke University Press, 2002), p. 30. [27–31]. Bazin's essay was originally published in the first version of his *What is Cinema? Volume 1* in 1958.

[40] Ranciére, p. 19.

Flusser, Vilém. *Towards a Philosophy of Photography.* Chicago: University of Chicago Press, 1983.

Förster, Annette. *Women in the Silent Cinema: Histories of Fame and Fate.* Amsterdam: Amsterdam University Press, 2017. https://doi.org/10.5117/978 9089647191.

Grecourt, Eugene. described the term as referring to "gangs of pimps and criminals" in *L'Intermediaire des Chercheurs et Curieux* 1032. 279 (29 Feb. 1904), p. NA.

Hudson, Dale. "Just Play Yourself, Maggie Cheung: Irma Vep, Rethinking Transnational Stardrom and Unthinking National Cinemas," *Screen* 47.2 (Summer 2006), pp. 213-232. https://doi.org/10.1093/screen/hjl017.

Kracauer, Siegfried. *Theory of Film: The Redemption of Physical Reality.* New York: Oxford University Press, 1960.

Marks, Laura U. *The Skin of the Film: Intercultural Cinema, Embodiment, and the Senses. Durham,* NC: Duke University Press, 2000.https://doi.org/10.23 07/j.ctv1198x4c.14.

McCauley, Scott. "Whenever Cinema is in Chaos, You Can Do a New Irma Vep": Olivier Assayas on His New HBO Series. Filmmaker.com (23 June, 2022). <<https://filmmakermagazine.com/115163-olivier-assayas-irma-vep-hbo/>> (Accessed Oct. 2022).

NA, Lettrist Cinema, <<https://www.spectacletheater.com/lettrist-cinema/>> (Accessed 11 Jan., 2024).

Picherit, Herve. "Making a Modernist Masterpiece: The Elusive Location of Cultural Significance in Les Vampires," *Modernist/modernity* 25.4 (November 2018), pp. 655-681. https://doi.org/10.1353/mod.2018.0054.

Ranciére, Jacques. *Emilian Battista,* trans. Film Fables. New York: Berg, 2006. https://doi.org/10.5040/9781350252066.

Chapter 8
Making Catastrophe Visible: Photography, Time, and Matter

Alina Cherry
Wayne State University

Abstract

This essay examines the role of analog photography in the aftermath of Japan's 3/11 disaster – an earthquake, tsunami, and nuclear accident – through emotional, temporal, and ethical lenses. It explores three key themes: the recovery of tsunami-damaged photographs, the visual representation of disaster, and the imaging of radioactive contamination. Drawing on theorists like Barthes, Sontag, and Berger, and artists such as Girard, Minato, Takeda, and Kawakubo, the essay shows how photography resists forgetting, fosters collective memory, and reckons with ecological peril.

The material fragility and persistence of analog photography afford it ontological depth, making it a powerful medium for remembrance and healing. Volunteer efforts to restore damaged images act as communal affirmations of life amid loss. Meanwhile, cameraless autoradiographic works – using contaminated soil to expose film – visualize invisible radiation and challenge human-centered perception. These images, shaped by nonhuman forces, serve as both ecological testimony and artistic resistance. Ultimately, the essay argues that the power of photography lies not only in documenting destruction but in revealing what catastrophe erases or renders invisible.

Keywords: Analog Photography, 3/11 Disaster (Japan), Collective Memory, Radiation Visualization, Ethics of Representation.

***

"No light, but rather darkness visible."

John Milton, *Paradise Lost*

Exploring the fate of photographs and the role of photography in the context of disaster affords remarkable insights into the reach and power of photography, particularly its ability to foster a sense of continuity in people's lives, aid in the restoration of communities, document loss, and serve as a witness for future generations. Photography illuminates, both literally and figuratively, the hidden and even invisible aspects of a disaster, shedding light on its most obscured dimensions and resisting the pull of oblivion. As Georges Didi-Huberman observes in his study of four photographs "snatched" at Auschwitz, "photography [...] shows a particular ability [...] to curb the fiercest will to obliterate" as it "works hand in glove with image and memory and therefore possesses their notable *epidemic power.*"[1] Importantly, examining the impact of surviving photographs in the aftermath of a disaster, alongside the significance and complexities of capturing such catastrophic events, reveals salient distinctions between analog and digital photography. In this essay, I propose to tease out several key functions of the photographic image within the context of the triple catastrophe that began to unfold in Japan on March 11, 2011: the earthquake, tsunami, and nuclear accident. A "compound disaster"[2] signaling the conjunction of natural and man-made elements, 3/11 engages with photography in complex ways not only due to the magnitude of the devastation it caused, but also, and especially, because of the challenges it poses to photography, and more extensively, to vision. This study unfolds like a triptych, addressing the impact, role, and contributions of photography through distinct yet interrelated instances: 1) the monumental effort to salvage, clean, and return irreplaceable family photographs to survivors in the wake of the tsunami; 2) the arguably impossible task of representing a catastrophe through photographic means; 3) the use of analog photography to show the invisible disaster unfolding at Fukushima, as the damaged reactors released radioactive particles into the atmosphere. While each section focuses on a different aspect of the disaster, they are bound together by the recurrence of analog photography as a meaningful player in the rebuilding efforts following the catastrophe, in philosophical discussions linking photography and catastrophe through the idea of technological progress, and in documenting "the violence of radiation ecologies."[3] Notably, each of the three segments offers unique perspectives on the ability of analog photography to operate as an art form that "make[s] visible" rather than

[1] Georges Didi-Huberman, *Images in Spite of All,* trans. Shane B. Lillis (Chicago and London: The University of Chicago Press, 2008), 23.

[2] Pablo Figueroa, "Subversion and Nostalgia in Art Photography of the Fukushima Nuclear Disaster," in *Fukushima and the Arts: Negotiating Nuclear Disaster,* ed. Barbara Geilhorn and Kristina Iwata-Weickgenannt (London: Routledge, 2017), 59.

[3] Elizabeth DeLoughrey, "Radiation Ecologies and the Wars of Light," *Modern Fiction Studies* 55, no. 3 (Fall 2009): 469.

merely "reproduce[s] the visible," as Paul Klee's well-known aphorism about the role of art suggests.[4]

Memories under Water

On March 11, 2011, at 14:46 Japan Standard Time, a magnitude 9.0 earthquake occurred in the Pacific Ocean near the northeast coast of Honshu, Japan's main island. It was the strongest earthquake ever recorded in Japan and the fourth largest worldwide since 1900. The tectonic activity triggered a tsunami with waves reaching nearly 130 feet, surging up to six miles inland in some areas. As they cascaded over seawalls, uprooted protective forests, and swept over coastal regions, destroying homes, businesses, and infrastructure across 1242 miles of coastline, the powerful tsunami waves also damaged the Fukushima Daiichi nuclear plant, causing the meltdown of three of its reactors. Consequently, toxic levels of radiation were spewed into the atmosphere, forcing the evacuation of approximately160,000 people.[5] The accident was classified as a category 7 event, the highest rating on the International Nuclear and Radiological Event Scale.[6]

The earthquake and tsunami produced 22.63 million tons of debris and waste. As water surged through hundreds of thousands of homes, personal belongings were swept away and merged into colossal debris pileups alongside tangles of metal, trees, remnants of buildings, damaged vehicles, and ships and boats washed inland. On March 25, the Japanese government made an official announcement calling for rescue teams, consisting of members of the Japan Self-Defense Forces and volunteers, to recover photographs and memorial items such as certificates, letters, and Buddhist tablets. As a result, numerous volunteer groups, referred to as PRPs (photography recovery projects), were quickly formed to clean tsunami-damaged photographs with the goal of returning them to their rightful owners. Photos were cleaned on-site when possible or sent to processing centers in other cities. In addition to these local initiatives, a large-scale campaign was launched by the Fuji Film corporation (The Fuji Film Photo Rescue project) after the company received countless

---

[4] Paul Klee, *The Thinking Eye: The Notebooks of Paul Klee*, ed. Jürg Spiller, trans. Ralph Manheim (London: Lund Humphries, 1961), 76.

[5] According to the official website of Fukushima prefecture: https://www.pref.fukushima.lg.jp/site/portal-english/en-1-4-1.html#:~:text=In%20May%202012%2C%20approximately%20160%2C000,individuals%20with%20unknown%20evacuation%20destinations.

[6] For reference, the 1986 accident at Chernobyl in the Ukrainian SSR (now Ukraine) was classified under the same category. Both Chernobyl and Fukushima are regarded as the most catastrophic nuclear accidents to date.

inquiries from volunteer groups. As part of this project, Fuji Film staff traveled to the affected areas to teach volunteers how to properly treat and clean soiled photographs.[7]

Yoko Shiraiwa, a photographic conservation specialist who volunteered in Ofunato, a coastal city in Iwate prefecture, details the painstaking labor required by this process:

> When the photographs were first delivered, they were dried by taking them out of the boxes they had been delivered in. Washing the photographs in water and drying them was the main treatment applied. However, it was not always so simple. Photographs that started to dissolve were difficult to wash. Where possible, the faces of people were left on the print so that it would be possible for the family to identify them. Blocks of photographs were immersed in water until they could be peeled off one by one. Pocket albums were cut along the rims to open them up and then placed in the water to remove the plastic film and paper. It was sometimes necessary to split magnetic albums so that the water could penetrate. Some magnetic albums could not be washed as the images transferred to the clear cover sheets; therefore, they were trimmed with the paper mount and cover sheet attached. Photographs stuck on to the glazing were extremely difficult to treat especially when the prints were on resin-coated paper which has polyethylene layers on both sides, making it impenetrable to water.[8]

Social anthropologist Fuyubi Nakamura, who participated in relief efforts in Miyagi prefecture, recalls a similar process. Both Shiraiwa and Nakamura highlight the urgency that drove the volunteers, as the photographs – already suffering considerable damage – were rapidly deteriorating. They also comment on the sense of purpose they found in this work during a time of national grief. Nakamura describes being "able to do something in the aftermath of the disaster" as "rewarding,"[9] while Shiraiwa insists on the collective sense of uplift

[7] In addition to the Fuji Film Photo Rescue project, the Tokyo Document Recovery Assistance Force (http://toubunq.blogspot.jp/) was established in June 2011 to support the emergency recovery system for documents and photographs.

[8] Yoko Shiraiwa, "Rescuing Tsunami-Damaged Photographs in Japan," *Journal of the Institute of Conservation* 36, no. 2 (2013): 198-99, https://doi.org/10.1080/19455224.2013.820205.

[9] Fuyubi Nakamura, "Memory in the Debris: The 3/11 Great East Japan Earthquake and Tsunami," *Anthropology Today* 28, no. 3 (2012): 7.

derived from cleaning hundreds of thousands of photographs: "it was something very optimistic to report from the affected areas."[10]

The concerted efforts to salvage photographs have also attracted the attention of writers. In *Fukushima: Story of a Disaster* (2012), a compelling chronicle of the triple calamity, Michaël Ferrier, a French academic and writer living and teaching in Tokyo, traveled to the disaster-stricken region in the wake of March 11. He describes the gestures of a volunteer, Rieko, who

> ranks all the documents patiently and meticulously. [...] But she is particularly careful with the photos. They have to be dried, cleaned, and organized. Clothespin, toothbrush, rags soaked in fresh water, she dabs gently. Little by little, the lines become lighter, the colors reappear. We feel that something other than simple cleaning is at stake. Gesture therapy. Through water, take back from water what water has taken from you. Rieko has it all figured out: she arranges, reframes, interprets.[11]

Rieko's repetitive gestures and simple body mechanics appear to have a curative quality, and as such, they constitute small steps in dealing with the widespread devastation and regaining a sense of normality. Ferrier illustrates how the collective act of gathering to clean photographs enables people to engage in conversation, reconnect socially, and rebuild communal ties. The photographs spark dialogue and trigger memories, allowing survivors to "recompose" their past. Ferrier maintains that "[F]or people who have lost everything, when their entire past has been reduced to dust or cut into pieces, a photograph means a lot."[12] It has been documented that, after the tsunami, photographs were the first items that people sought to recover.[13] The large number of volunteer groups and the strategic involvement of the Fuji Film corporation underscore the importance of these family heirlooms for catastrophe survivors, while inviting reflection on why photographs hold more significance than other personal possessions. Can the value attributed to photographs be explained through their complex ontology?

In "Rhetoric of the Image," an article he penned for the journal *Communications* in 1964, years before his seminal work *Camera Lucida* (1980), Roland Barthes argues that the ontological specificity of the photograph lies in its ability to

---

[10] Shiraiwa, 201-02.

[11] Michaël Ferrier, *Fukushima: récit d'un désastre*, my translation (Paris: Gallimard, 2012), 210-11.

[12] Ferrier, *Fukushima*, 208-09.

[13] Elissa Curtis, interview with Munemasa Takahashi, "Lost & Found: Salvaging Snapshots in Japan," *The New Yorker*, March 19, 2012.

create a new space-time category, rendered through the concept of *having-been-there.* This concept simultaneously designates spatial immediacy and temporal anteriority. The photograph is experienced in the *here-now* as a reality that once existed but is no longer. The photo thus provides "stupefying evidence of *this is how it was,* giving us, by a precious miracle, a reality from which we are sheltered."[14] In the post-tsunami context in Japan, for a survivor who has endured incredible human and material loss, a recovered photograph – even if damaged – may be the sole remaining evidence of a past life, affirming that a certain reality once existed. It is intriguing that Barthes speaks of this past reality in terms of something to be "sheltered" from ("à l'abri" in French), whereas in post-disaster Japan, the relationship between the viewer and the reality captured by the photograph was reversed; the past reality of the photograph provided, in fact, a refuge from the immediate, unbearable reality of the catastrophe.

The damage to the photographs varied widely; while some showed only minimal scarring, others were partially or entirely effaced. Shiraiwa recalls that most of the photographs she worked on were printed after the 1970s and were chromogenic color prints that used either fiber-based or resin-coated paper as their support.[15] When looking over rescued photographs, we notice how the contaminated tsunami water ate away, like a virus, at the support of the image, dissolving the gelatin layer and attacking the paper itself. The disaster revealed a side of the photograph that we, as viewers who typically marvel at the finished product, are not usually privy to: the photo as a chemical process. While vulnerable, as their disfigurement attests, the fragile materiality of these analog pictures is precisely what enabled them to endure, "as none of digital photographs stored on hard disks, CDs or memory sticks survived. [...] Photographs thus became objects of memory and relics at the same time."[16] The recovered photographs were eventually digitized to facilitate their circulation and potential recognition by their owners. Yet, the fact that such a large number of prints were reclaimed by survivors evinces their attachment to analog photographs as tangible artifacts infused with transgenerational presence and family history. These photographs, then, are more than "significant surfaces"[17]; they are cherished items that have been touched, organized, and preserved by family members. In

[14] Roland Barthes, "Rhetoric of the Image," in *Image, Music, Text,* trans. Stephen Heath (New York: Hill and Wang, 1977), 44.

[15] Shiraiwa, 197.

[16] Nakamura, 7.

[17] Vilém Flusser, *Towards a Philosophy of Photography* (London: Reaktion Books, 2000), "The Image," Kindle, n.p.

contrast to analog images, digital photographs are "less proximate, less tangible, less possessable"; furthermore, the screen alters "the effectiveness of the photograph as a vehicle of memory."[18] This shift is especially significant in the context of a disaster, where loss can only be mitigated through acts and objects of remembrance.

Fig. 8.1: Lost & Found Exhibit 1/11-2/11 2012, AKAAKA Gallery, Tokyo, Japan.

On these weathered surfaces, we encounter fragmented realities and erased identities. Unlike the intact photographs Barthes analyses in *Camera Lucida*, which lead him to assert that "a photograph is always invisible: it is not it that we see,"[19] the rescued photographs become visible in themselves, their materiality violently exposed by the damage they have sustained. Human faces are partially rubbed out or have completely vanished, some colors have faded, while others remain eerily vivid. In certain photographs, the paper bears traces of mold or has lost its smoothness, appearing creased or dull. A number of pictures only display remnants of color – yellows, blues, reds, maroons, browns – in swirls, splotches, and geometric shapes, evoking an abstract expressionist style, tsunami Pollocks. When still visible, the scenes in the photographs capture moments of everyday life in its natural simplicity: a ski trip, two children playing in a pink

[18] Emily Keightley and Michael Pickering, "Technologies of Memory: Practices of Remembering in Analogue and Digital Photography," *New Media and Society* 16, no. 4 (2014): 9.

[19] Roland Barthes, *Camera Lucida*, trans. Richard Howard (New York: Hill and Wang, 1981), 6.

pool, home and travel scenes, portraits, a class picture, and so on. Yet, entire sections of these images have disappeared, leaving the whitish surface beneath exposed. The latter is not meant to be seen, its presence usually concealed by the image of the photograph; when laid bare, as in the case of the ruined pictures, it paradoxically constitutes an excess, the very excess at the core of Maurice Blanchot's definition of disaster: "The disaster would be in addition, in excess, an excess which is marked only as impure loss."[20] Moreover, by whitening out parts of the photograph, the disaster has not only altered the image and its support, but also its temporality. On the original instant frozen in time by the camera, the disaster has superimposed its own chronology: the arrival of the tsunami wave, and the varying lengths of time the photographs spent in contact with polluted water and debris. Each degree of damage reflects this passage of time. Even if these photographs do not depict the catastrophe, they "make it visible" through their deteriorated materiality and layered temporality, allowing us to visualize the disaster as emerging from within their altered form.

As previously stated, the photography recovery projects were started with the goal of returning the photographs to their owners. Shiraiawa reports that from the beginning of June 2011 to the end of June 2012, her volunteer group cleaned 263,000 photographs, of which 196,000 were returned to their owners.[21] Another volunteer group, Memory Salvage, digitized over 750,000 images, with more than 460,000 successfully returned to their owners. The Memory Salvage volunteers placed the photographs that hadn't been claimed (either because their owners had disappeared in the disaster or because they were too damaged to be recognized) in a "hopeless box." Munemasa Takahashi, a photographer working with Memory Salvage, had the idea of curating these photographs. His initiative gave birth to the Lost & Found project, a traveling exhibition (held in Japan, the US, Australia, and Europe) aimed at raising funds for tsunami survivors.[22] The hesitation to discard these abandoned photographs is indicative of their compelling presence as residues of the past – indexical traces of the presence of their subjects. While we may question the displacement of their intended purpose and the ethics of their passage from the private to the public realm, from family collections to public exhibitions, the cumulative effect of over 3,000 tsunami-damaged photographs powerfully conveys both the testimonial

---

[20] Maurice Blanchot, *The Writing of Disaster*, trans. Ann Smock (Lincoln: University of Nebraska Press, 1995), 6.

[21] Shiraiawa, 200.

[22] More information as well as some of the photographs included in the exhibition can be found on the website of the *Lost & Found* project: http://lostandfound311.jp/en/.

value of photography and the physical and emotional trauma caused by the disaster.

Fig. 8.2: *Lost & Found* Exhibit 3/8-3/25 2012, Hiroshi Watanabe Studio, Los Angeles, US.

In an essay dedicated to the *Lost & Found* exhibit, Geoffrey Batchen reflects on the impact of this photographic assemblage:

> [...] the guilty pleasure we feel before such beauty is only possible when we look at these images *en masse*, when we can repress the individual pictures and see only the installation. In that situation, photographs become "photography," with the inverted commas a rhetorical invitation to forget the specifics of events in Japan (forget why these photographs are here, forget the 19,000 deaths and the millions who are still struggling to recover) and consider other possible ways to regard this ensemble of pictures.[23]

The installation, Batchen suggests, can be viewed as an expression of our desire to photograph and be photographed, a desire that, in itself, reflects our need to transcend death by "providing witness to existence in general."[24] Significantly, Batchen mourns a double loss as he draws a parallel between the

---

[23] Geoffrey Batchen, "Reverie: Lost and Found," in *Tsunami, Photographs, and Then* (Tokyo: AKAAKA, 2014), 142, http://lostandfound311.jp/en/essay/index.html.

[24] Batchen, 142.

damage sustained by these photographs and the disappearance of the analog mode of representation from which they stem, a method now swept away by a "digital, rather than oceanic, tsunami."[25]

The Challenges of Making Disaster Visible

In the wake of the triple disaster in Japan, countless photographs taken by both professional and amateur photographers circulated widely in the media and on the internet. Most of these images focused on the tsunami and its aftermath, capturing scenes of wreckage, obliterated urban areas, and scarred landscapes. Commentators and viewers commonly referred to them as disaster photography. But what does this designation truly imply, and what do these photographs actually capture? Is seeing the effects of a disaster, or an image of a colossal tsunami wave crashing onto the shore, equivalent to seeing the disaster itself? To what extent can a catastrophe be represented through photography? To grapple with these questions, I turn to a range of perspectives – some theoretical, others practice-based, voiced by visual artists who have documented disasters – that discuss and even challenge the possibility of representing such events through photographic means.

In her article "Photographs of Disasters. An Ethnomethodological Approach," sociologist Ruth Ayaβ investigates disaster photography through a series of photographs produced and disseminated during and after the March 2011 disasters in Japan. While her argument centers on how disaster photographs show the destruction of order by instituting a new type of visual order, and how these images of ruin and suffering evoke pain in the viewer, she also discusses the rather impossible task faced by photographers in attempting to capture a catastrophe as it unfolds. Unlike war photography, which can be viewed as a type of disaster photography where photographers can get very close to the event (as war photographer Robert Capa famously said, "if your pictures aren't good, you aren't close enough"[26]), disaster photographs are generally taken in the aftermath of the event. For a photograph to exist, a photographer must be present at the exact moment an event unfolds, and since most disasters are unpredictable, there is often a delay between the occurrence of the disaster and the moment the photograph is taken. These delays are also due to the dangers that catastrophes pose to photographers and the challenges of accessing devastated areas. However, the emergence of social media and mobile phone cameras have considerably facilitated the capture and circulation of disaster

[25] Batchen, 142.

[26] Ruth Ayaβ, "Photographs of Disasters. An Ethnomethodological Approach," *Visual Studies* 35, no. 2-3 (2020): 171, https://doi.org/10.1080/1472586X.2020.1775494.

images. Even professional journalists now rely, in their reporting, on images taken by amateur photographers using mobile devices.[27] And yet, while some disasters, such as Hurricane Katrina, can be photographed as they approach, others, like the earthquake that occurred off the northeastern coast of Honshu at a depth of 18.6 miles below the Pacific Ocean floor, elude photographic capture.[28] Ayaβ identifies the sheer size of disasters as another important barrier to their visual representation. For instance, the arrival of the tsunami wave in Japan was captured by multiple cameras installed by cities in key coastal areas, but these images represent only segments of the gigantic tidal wave: "The impossibility of depicting the entirety of a disaster [...] bears witness to its unfathomable scale in that it resists visual representation."[29]

One could argue that this partitioning of reality is not unique to disaster photography, but rather intrinsic to what photography is and does. "A photograph [...] isolates, preserves, and presents a moment taken from a continuum,"[30] asserts John Berger. Susan Sontag also speaks of photography as omission, famously stating, "to frame is to exclude"[31] in her landmark study *Regarding the Pain of Others.* Both Berger and Sontag highlight, albeit in different ways, the dynamic that photography institutes between what is shown and what is not shown, between the visible and the invisible – in short, between presence and absence. Whereas Sontag associates exclusion with the photographer's decision to focus on a particular content at the expense of others, Berger defines it in temporal terms: "This choice is not between photographing X and Y: but between photographing at X moment or at Y moment."[32] For him, the "true content of a photograph is invisible, for it derives from a play, not with form, but with time."[33] Thus, what is visually absent from a photograph is determined by the photographer's temporal choices and ultimately invoked by what is shown.

The relationship between photography and time is particularly significant in disaster photography, not only due to the delays often imposed by the event itself, as Ayaβ remarked, but also because of the representational limits of events that require time to be processed, since catastrophes tend to overwhelm our capacity to fully grasp them in the moment. A distinction must therefore be

---

27 Ayaβ, 171.

28 Ayaβ, 175.

29 Ayaβ, 177.

30 John Berger, *Understanding a Photograph* (New York: Aperture, 2013), 26.

31 Susan Sontag, *Regarding the Pain of Others* (New York: Picador, 2003), 46.

32 Berger, 26.

33 Berger, 25.

made between photography that merely records the disaster in its immediate aftermath and photography that aims to represent its deeper implications. In a media landscape that facilitates the instantaneous distribution of images by both amateur and professional photographers, a number of visual media artists deliberately chose to pause in order to reflect on the complexities of representing a catastrophe and on what it means "to make it visible." Among them is Thierry Girard, who devised a temporal framework for photographing disasters, dividing the process into three distinct stages.[34] The first stage is defined by the very moment of the catastrophe itself. Girard brings up major disasters such as Christchurch, Fukushima, and 9/11, all of which have received extensive media and photographic coverage. He argues that photographers working in the immediate aftermath of such events face a sensory overload that gives rise to a paradox: "the more intense and violent the event, the blinder the operator."[35] This excess of sensory input leaves the photographer "in a state of absolute bewilderment," confronted with what he calls "the visual unthought" ("l'impensé visuel").[36] In this state of astonishment, which Girard equates with blindness, the photographer becomes nothing more than a recorder of the disaster, lacking the psychological and intellectual distance needed "to face"[37] the event. The effect of this astonishment is such that it "creates a form of helplessness of the immediate gaze."[38] Therefore, in order to make the catastrophe visible to an audience, the photographer must find this critical distance without which they are a mere operator, passively recording a calamity.

The second and third phases that Girard proposes are designed to avoid this pitfall. The second stage implies the passage of time and only takes place months, or even years, after the catastrophe. Girard practiced this approach at Kamaishi, a coastal city severely damaged by the tsunami, visiting the site twice, once five months after the disaster, and again eighteen months later. By embracing the passage of time, Girard allowed for the progressive erasure of the disaster's traces, which in turn enabled him to problematize its representation and search for "a threshold, an in-between" – a moment of hesitation and questioning of what can precisely be crossed or overcome.[39] The third stage unfolds over an

---

[34] Thierry Girard, Hélène Lucien, Minato Chihiro, Marc Pallain, and Michaël Ferrier, "Donner à voir Fukushima," in *Dans l'œil du désastre: créer avec Fukushima*, ed. Michaël Ferrier (Vincennes: Thierry Marchaisse, 2021), 139.

[35] Girard et al., 140.

[36] Girard et al., 140.

[37] Girard et al., 140.

[38] Girard et al., 140.

[39] Girard et al., 143.

even longer time span and is predicated on the photographer's visit to a disaster site decades, or even centuries, after the event. In this case, its traces are no longer visible but layered within the history of the landscape, waiting to be deciphered – like a "palimpsest"[40] – by the photographer's attentive eye. Hence, to reveal the catastrophe, Girard seeks a distanced perspective, an elsewhere that is not spatial but temporal in nature.

While the first stage confronts the photographer with an excess that is difficult to process sensorily, intellectually, and emotionally, in the second and third stages of representation proposed by Girard, the photographer is no longer concerned with capturing traces of the event; instead, they focus on erasure and forgetting. In this context, photography becomes "a subtraction exercise."[41] The photographer redirects attention from the event itself to its location, abandoning the immediate gaze in favor of a temporally remote one. The landscape then becomes both a guide and a tool, enabling the photographer to access the embedded history of the place.

Nobuhiro Suwa, a Japanese film director, echoes the idea of temporal distancing as essential for an artist to make a catastrophe visible. He waited almost a decade before filming *Kaze no denwa* (*Voices in the Wind*, 2020) in tsunami-damaged areas. Suwa argues that immediately after the disaster, it would have been impossible to capture the event meaningfully, as the traces of destruction on the landscape would have appeared banal.[42] Suwa's stance is not surprising, given the exorbitant number of images that flooded the media and the internet in the immediate aftermath of the disaster. Additional images wouldn't elicit a different reaction from the viewer; rather, they would have what Sontag refers to as a "diminishing effect,"[43] causing the event to appear increasingly less real, and contributing to the viewer's desensitization. By waiting to photograph, Girard and Suwa, alongside many other Japanese artists working in photographic media, adopted an "anti-reportage stance," which entailed "step[ing] back from simply recording the disaster to reflect on the larger conceptual issues raised by creating art from such harrowing events."[44]

---

[40] Girard et al., 143.

[41] Girard et al., 142.

[42] Nobuhiro Suwa and Élise Domenach, "Filmer ce qu'on ne peut pas filmer," in *Dans l'œil du désastre: créer avec Fukushima*, ed. Michaël Ferrier (Vincennes: Thierry Marchaisse, 2021), 189.

[43] Sontag, 105.

[44] Anne Nishimura Morse and Anne E. Havinga, eds., *In the Wake: Japanese Photographers Respond to 3/11*, (Boston: MFA Publications, 2015), 12.

Minato Chihiro, a Japanese photographer, arts curator and art theorist, defines the photographic act along similar lines of economy, subtraction, and even failure. "To take a photograph is to miss,"[45] he asserts. His reflections on photography are intertwined with a questioning of both human and technological limits, shaped by his experiences as a photographer in the village of Iitate, in Fukushima prefecture. Witnessing nature flourishing wildly in what has become a ghost town after the evacuation of inhabitants following the nuclear accident, he observes that the limits imposed on human habitation have paradoxically led to the thriving of the habitat. The abandonment of the village, he reflects, was itself the result of crossing a technological threshold – one initiated by the splitting of the atom and the arrival of the atomic age.[46] With this observation, Chihiro prompts us to ruminate on the limits of technology and their implications for humankind. As concerns photography, he views technological progress with skepticism, because the advent of digital photography and mobile phones has made it increasingly difficult to fail in capturing a picture. He reminds us that for more than a century, taking a bad photograph was commonplace due to the imprecision of earlier cameras. Yet, by perfecting the digital process, we have reached a technological limit, Chihiro argues, connecting this development to the nuclear age through the idea of visibility. What happened at Fukushima, he suggests, places limits on what can be visualized, since radiation is invisible to our naked eye and impossible to photograph (but not impossible to show through camera-less techniques, which I address next). Fukushima, then, represents both a warning against technological hubris and a stark reminder of our human fallibility.

Fukushima: Capturing the Invisible

As we have noted, 3/11 challenges not only representation but visibility itself. While the impact of the earthquake, tsunami, and explosions at the Fukushima Daiitchi plant could be seen, the radioactive contamination from nuclear particles released into the air remained imperceptible to the human eye. Nonetheless, radiation can be revealed through analog photography methods, as "celluloid film is sensitive to environments operating beyond the human

---

[45] Girard et al., 145.

[46] A crossing of limits is what characterizes catastrophe itself, according to philosopher Jean-Luc Nancy who draws comparisons between Auschwitz and Hiroshima, and between Hiroshima and Fukushima: "What is common to both these names, Auschwitz and Hiroshima, is a crossing of limits – not the limits of morality, or of politics, or of human dignity, but the limits of existence and of a world where humanity exists, that is, where it can risk sketching out, giving shape to meaning." Jean-Luc Nancy, *After Fukushima: The Equivalency of Catastrophes*, trans. Charlotte Mandell (New York: Fordham University Press, 2015), 12.

senses."[47] Photography and radiation have, in fact, a long-shared history that dates back to the last decade of the nineteenth century. After German physicist Wilhelm Röntgen's accidental discovery of X-rays in 1895, while working with fluorescent light, French physicist Henri Becquerel began experiments to understand why the minerals used for X-rays exhibited luminescence.[48] In 1896, Becquerel made an accidental discovery of his own when he exposed fluorescent uranium salts to Lumière Etiquette Bleue glass photographic plates covered in black paper.[49] He noticed that an exposure had formed without the help of visible light, "as though the earthly matter itself was reaching out and participating in photographic processes with a determined non-human agency."[50] Becquerel's experiments led to the discovery of radiation, and the relationship between photography and radiation has since intrigued both amateur enthusiasts and artists like elin o'Hara slavick, Julian Charrière, Giles Price, Marc Pallain, Hélène Lucien, and Masamichi Kagaya. In this section, I discuss works by two Japanese artists, Shimpei Takeda and Yoi Kawakubo, who, in response to the 3/11 disaster, have experimented with the effects of radiation on light-sensitive film to render this invisible danger visible, thereby drawing attention to the ecological crisis unfolding in Japan and its long-term consequences.

Both Takeda and Kawakubo have used contaminated soil "to complete the iconography of the nuclear accident by providing an informative or reflexive dimension that had been set aside by the mass media."[51] The use of soil itself carries powerful symbolism, as land is what allows humans to dwell, build homes, and form communities. It is also what sustains life in all its forms. When land becomes contaminated, it is rendered uninhabitable, triggering mass evacuations, and leading to exile and exclusion. This symbolism is evident in how Takeda conceptualized his "Trace" series. For this project, Takeda and his crew collected 16 soil samples from 12 locations across five different prefectures. The sites were selected specifically for their ability to "contain a strong memory of life and death": temples, shrines, war sites, ruins of castles, and Takeda's birthplace in Fukushima prefecture.[52] The artist admits to having an "odd and

---

[47] Thomas Pringle, "Photographed by the Earth: War and Media in Light of Nuclear Events," *Necsus: European Journal of Media Studies* 3, no. 2 (2014): 137.

[48] Pringle, 136.

[49] Pringle, 136.

[50] Pringle, 136.

[51] Amandine Davre, "Revealing the Radioactive Contamination after Fukushima in Japanese Photography," *Trans Asia Photography Review* 10, no. 1 (2019): n.p.

[52]Shimpei Takeda, "Trace – Cameraless Records of Radioactive Contamination," in *Making the Geologic Now*, ed. Elizabeth Ellsworth and Jamie Kruse (New York: punctum books, 2013), 210.

terrifying experience" during the collection process, noting that "the reading on the radiation meter climbed uncomfortably," and he felt as though he were "gathering somebody's ashes."[53] Takeda viewed these soil-collecting trips as a pilgrimage, a tribute to his deep-rooted connection with the land where his ancestors were buried, the same land now contaminated by nuclear fallout.[54] The title of the series can thus be interpreted as an attempt to capture both the traces of the present, in the form of radiation, and those of the past, as soil was collected from locations tied to Japan's history.

Takeda stored the soil samples alongside 8x10 B/W sheet films in light-tight containers for a month, allowing the radiation to affect the film. He then developed the film, obtaining autoradiographs through a process that he terms "cameraless."[55]

Fig. 8.3: Shimpei Takeda, Trace #7, Nihonmatsu Castle, from the *Trace* series, 2012, 50.8 x 61 cm., gelatin silver print.

[53] The photographer's powerful confession can be read in its entirety on his website: https://www.shimpeitakeda.com/trace/note/.

[54] Takeda, 209.

[55] Takeda, 206.

The resulting images evoke the appearance of stars, nebulae, and constellations. In fact, an uninformed viewer might innocently mistake the harmful radiation for a spectacular view of the night sky. Notably, disaster is deeply embedded in Takeda's "starry night," first through its toxic origins, and second, metaphorically, as the etymology of the word "disaster" itself suggests. Derived from the Italian *disastro*, it refers to an ill-starred event caused by the unfavorable position of a planet or star, or the state of being forsaken by a protective celestial body. Moreover, the cosmic night fashioned through the "Earth's autonomous aesthetic practices"[56] recalls the supernova explosions from over 6 billion years ago, events that produced uranium. As such, it tells a story of both destruction and creation, primarily the birth of heavy elements like uranium, which later played a crucial role in the formation of the Earth. The darkness in the image is also evocative of the origins of photography, which, in its early days, was known as "the black art"[57] due to the collodion that would stain the photographer's fingers.

Unlike Takeda, Yoi Kawakubo came to discover the effects of radiation on film by accident. During a yearlong artist residency at Tokyo Wonder Site, Kawakubo photographed the landscape in the Fukushima exclusion zone and buried the film there. When he developed the film three months later, he noticed the photographs had an "unusual" low contrast, as if the film had been accidentally exposed to weak light before development.[58] He then started to experiment with Polaroids and realized that the film reacted directly to the radioactivity emanating from the soil. For his series of five large-format prints titled *If the radiance of a thousand suns were to burst at once into the skies* (2014-), Kawakubo buried unexposed silver halide film in different locations within the exclusion zone in September 2013 and retrieved it in the spring of 2014. The title of the series highlights a connection to Japan's more recent past, specifically the atomic bombings of Hiroshima and Nagasaki during WWII, events that permanently reshaped the relationship between humans and their environment. Kawakubo's title is a verse from the renowned Hindu scripture, the Bhagavad Gita, which was famously quoted by Robert Oppenheimer, the American physicist who oversaw the development of the first atomic bomb, upon witnessing the first atomic explosion at the Trinity site in New Mexico. By choosing this title, Kawakubo aimed to link a local event to the age-old human desire for power, the history of nuclear energy, and the drive for productivity that defined the twentieth century, seeking "to show the compromise facing

[56] Pringle, 149.

[57] Carol Mavor, *Black and Blue* (Durham: Duke University Press, 2012), 42.

[58] Davre, n.p.

our contemporary way of life."[59] The print in Fig. 4 seems to encapsulate the title of the series through an explosion of yellows and oranges varying in shade and saturation. As the colors intensify towards the center, they point to an incandescence reminiscent of the sun's duality: a life-sustaining force and a potentially lethal energy. Sunlight is also indispensable to photography, yet Kawakubo's pictures are, paradoxically, produced underground, relying on another energy source that is not naturally occurring but man-made.

Fig. 8.4: Yoi Kawakubo, *If the radiance of a thousand suns were to burst at once into the skies, I*, 2014.

The photo in Fig. 5 captures not only the effects of radiation but also traces of mold that formed inside the film holder while it was buried. Radiation cannot be isolated from the natural elements affecting the soil – such as mold, moisture, and weather variations – in the formation of the image. The photographer plays only a partial role in this operation, neither framing, nor shooting, only providing the materials for the radioactive soil to generate a photographic image, much like a "self-portrait,"[60] and ultimately developing

[59] Yoi Kawakubo, "L'art peut sauver le monde," in *Dans l'œil du désastre: créer avec Fukushima*, ed. Michaël Ferrier (Vincennes: Thierry Marchaisse, 2021), 81.

[60] I am borrowing the term used by Kate Brown to describe traces of radioactive pollution captured in photographs taken at Chernobyl: "These points of light are not representations.

the film. The notion of a self-portrait is particularly fitting, as the long exposure time renders this process more akin to painting than to the instantaneous nature of contemporary photography. By revealing a process through which the earth itself becomes the artist, one that "recognizes the powerful, invisible authorship of radioactivity,"[61] these photographs also raise politico-ethical questions about human involvement in the alteration of the environment. They urge us to ponder the weight of our responsibility in a universe where humans, intertwined with other non-human species, forces, and agents, are both shaped by and capable of shaping the delicate balance of harmonious coexistence.

Fig. 8.5: Yoi Kawakubo, *If the radiance of a thousand suns were to burst at once into the skies, V,* 2019.

They are energy embodied. The specks are none other than cesium, plutonium, and uranium self-portraits." Kate Brown, "Marie Curie's Fingerprint: Nuclear Spelunking in the Chernobyl Zone," in *Arts of Living on a Damaged Planet,* eds. Anna Lowenhaupt Tsing, Heather Swanson, Elaine Gan, and Niels Bubandt (Minneapolis: University of Minnesota Press, 2017), 41.

[61] Robert MacFarlane, "Response by Dr. Robert MacFarlane," *Picturing the Invisible,* n.d., accessed December 3, 2024, https://picturing-the-invisible.art/works/02-if-the-radiance-of-a-thousand-suns-were-to-burst-at-once-into-the-sky.

In helping create these images of environmental toxicity, Takeda and Kawakubo sought to give concrete form to the esoteric scientific discourse and the abstract radiation figures released daily by the Japanese government in the wake of the Fukushima accident.[62] While the artists' intentions seem relatively straightforward, situating them aesthetically presents a greater challenge. The post-nuclear photographic act, deeply enmeshed in a destructive light ecology, inaugurates a visual order that aims to reveal to the naked eye the surreptitious forces that can harm it. The autoradiographs cannot be entirely classified as "technical images," [63] as the latter are produced by apparatuses, whereas the autoradiographs are made without the use of a camera. However, both the film and the developing process, are products of technology. Takeda and Kawakubo's photographs can also be likened to what Joanna Zylinska terms "nonhuman photography," which she defines as a set of imaging practices where the human is absent, both as subject and agent, thus "expand[ing] the human-centric concept of photography."[64] Yet, Zylinska broadens the scope of nonhuman photography to photographs that are not created by humans, such as those produced by traffic-control cameras, microphotography and Google Street View, to include photographs that are not intended for humans (such as QR codes), and uncanny-looking images that are not of humans (e.g., depopulated sweeping landscapes). Importantly, Zylinska's project underscores the need for a new conceptual framework for understanding photography, one that transcends a human-centric perspective and resonates with posthumanist media theory. Media scholar Thomas Pringle also addresses this shift, employing terms such as "Earthly photography" and "radiation aesthetics," to describe art produced by energy processes "operating beyond anthropocentric intent."[65] In a manner similar to Zylinska, Pringle advocates moving beyond our "anthropocentric reliance on human perception in environmental representation."[66] In a bold conceptual move, philosopher Hervé Couchot proposes the category of "the inaesthetic"[67] to designate artworks authored by radioactivity. He draws on the Greek etymology of the term "aesthetic," which foregrounds sensory perception, in developing this framework. In so doing, he repurposes the concept of the "inaesthetic" to signal a shift from indifference to Beauty and the impossibility

---

62 Davre, n.p.

63 Flusser, "The Technical Image," Kindle, n.p.

64 Joanna Zylinska, *Nonhuman Photography* (Cambridge, MA: MIT Press 2017), 5.

65 Pringle, 149.

66 Pringle, 134.

67 Hervé Couchot, "L'inesthétique," in *Dans l'œil du désastre: créer avec Fukushima*, ed. Michaël Ferrier (Vincennes: Thierry Marchaisse, 2021), 239.

of creating works of art, toward modes of exhibiting that very impossibility, thus moving from negative aesthetics to aesthetic manifestations shaped by forces that elude the senses.[68]

Through analyses of photographs that do not directly represent the disaster but covey it through the material degradation of the image and its support, alongside critical debates on the representational and aesthetic challenges of capturing catastrophe on camera, and the examination of photographic practices that have the capacity to reveal the invisible phenomenon of radioactivity, I have explored some of the intricate modalities through which analog photography mediates and renders disaster visible. This study has also revealed the crucial connection between photography and time, first, in how disaster imposes its own chronology onto extant photographs through damage that both renders the event visible and introduces a forced temporality that disrupts and reconfigures the original temporal structure of the photograph. Second, as both professional and amateur photographers grapple with challenges of scale and accessibility in documenting the event, a number of media artists embrace the concept of temporal distance as a means of fostering a more nuanced and reflective engagement with the site of the disaster. Third, the process of revealing the ecological catastrophe unfolding in and around Fukushima through analog photography methods inherently involves time, as the film must undergo a prolonged period of cohabitation with radioactive soil in order for the particles to imprint their traces on the photographic surface. At the same time, the question of temporal scale inevitably arises for both the photographer and the viewer, considering that radioactive pollution is a long-term phenomenon, its presence measured not in years or decades, but in centuries and even millennia, with certain isotopes remaining active and potentially harmful for generations. The case studies, along with the theoretical insights presented, underscore the vital role of analog photography not only in documenting loss and bearing witnessing for future generations, but also in repairing the fabric of fractured communities. Not only does photography work as a social bonding agent, fostering connection and remembrance, but through its creative approach to catastrophe, it unveils overlooked disaster sites and ecocidal realities that evade conventional media frameworks.

Works Cited

Ayaβ, Ruth. "Photographs of Disasters: An Ethnomethodological Approach." *Visual Studies* 35, no. 2-3 (2020): 169-192. https://doi.org/10.1080/1472586X.2020.1775494.

[68] Couchot, 239.

Barthes, Roland. "Rhetoric of the Image." In *Image, Music, Text*, translated by Stephen Heath. New York: Hill and Wang, 1977.

Barthes, Roland. *Camera Lucida*. Translated by Richard Howard. Hill and Wang, 1981.

Batchen, Geoffrey. "Reverie: Lost and Found." In *Tsunami, Photographs, and Then*, edited by Munemasa Takahashi. Tokyo: AKAAKA, 2014. http://lostandfound311.jp/en/essay/index.html.

Berger, John. *Understanding a Photograph*. New York: Aperture, 2013.

Blanchot, Maurice. *The Writing of Disaster*. Translated by Ann Smock. Lincoln: University of Nebraska Press, 1995.

Brown, Kate. "Marie Curie's Fingerprint: Nuclear Spelunking in the Chernobyl Zone." In *Arts of Living on a Damaged Planet*, edited by Anna Lowenhaupt Tsing, Heather Swanson, Elaine Gan, and Niels Bubandt. Minneapolis: University of Minnesota Press, 2017.

Couchot, Hervé. 2021. "L'inesthétique." In *Dans l'œil du désastre, créer avec Fukushima*, edited by Michaël Ferrier. Vincennes: Thierry Marchaisse.

Curtis, Elissa. "Lost & Found: Salvaging Snapshots in Japan." *The New Yorker*, March 19, 2012. https://www.newyorker.com/culture/photo-booth/lost-found-salvaging-snapshots-in-japan.

Davre, Amandine. "Revealing the Radioactive Contamination after Fukushima in Japanese Photography." *Trans Asia Photography Review* 10, no. 1 (2019): n.p. https://doi.org/10.1215/215820251_10-1-104.

DeLoughrey, Elizabeth. "Radiation Ecologies and the Wars of Light." *Modern Fiction Studies* 55, no. 3 (Fall 2009): 468-495. DOI: 10.1353/mfs.0.1630.

Didi-Huberman, Georges. *Images in Spite of All*. Translated by Shane B. Lillis. Chicago: The University of Chicago Press, 2008.

Ferrier, Michaël. *Fukushima: récit d'un désastre*. Paris: Gallimard, 2012.

Figueroa, Pablo. "Subversion and Nostalgia in Art Photography of the Fukushima Nuclear Disaster." In *Fukushima and the Arts: Negotiating Nuclear Disaster*, edited by Barbara Geilhorn and Kristina Iwata-Weickgenannt. London: Routledge, 2017.

Flusser, Vilém. *Towards a Philosophy of Photography*. Reaktion Books, 2000. Kindle.

Fukushima Prefectural Government. "Transition of the Number of Evacuees." *Fukushima Revitalization Information Portal Website*. Accessed September 12, 2025. https://www.pref.fukushima.lg.jp/site/portal-english/en-1-4-1.html#:~:text=In%20May%202012%2C%20approximately%20160%2C000,individuals%20with%20unknown%20evacuation%20destinations.

Girard, Thierry, Hélène Lucien, Minato Chihiro, Marc Pallain, and Michaël Ferrier. "Donner à voir Fukushima." In *Dans l'œil du désastre: créer avec Fukushima*, edited by Michaël Ferrier. Vincennes: Thierry Marchaisse, 2021.

Kawakubo, Yoi. "L'art peut sauver le monde." In *Dans l'œil du désastre: créer avec Fukushima*, edited by Michaël Ferrier. Vincennes: Thierry Marchaisse, 2021.

Keightley, Emily, and Michael Pickering. "Technologies of Memory: Practices of Remembering in Analogue and Digital Photography." *New Media and Society* 16, no. 4 (2014): 576-593. https://doi.org/10.1177/1461444814532062.

Klee, Paul. *The Thinking Eye: The Notebooks of Paul Klee.* Edited by Jürg Spiller. Translated by Ralph Manheim. London: Lund Humphries, 1961.

Lost & Found Project. *Lost & Found Project (website).* Accessed September 12, 2025. http://lostandfound311.jp/en/.

MacFarlane, Robert. "Response by Dr. Robert MacFarlane." *Picturing the Invisible,* n.d. Accessed December 3, 2024. https://picturing-the-invisible.art/works/02-if-the-radiance-of-a-thousand-suns-were-to-burst-at-once-into-the-sky.

Mavor, Carol. *Black and Blue.* Duke University Press, 2012.

Milton, John. *Paradise Lost.* Hackett Publishing Company, Inc., 2005. Kindle.

Morse, Anne Nishimura, and Anne E. Havinga, eds. *In the Wake: Japanese Photographers Respond to 3/11.* Boston: MFA Publications, 2015.

Nakamura, Fuyubi. "Memory in the Debris: The 3/11 Great East Japan Earthquake and Tsunami." *Anthropology Today* 28, no. 3 (2012): 20-23. http://onlinelibrary.wiley.com/doi/10.1111/j.1467-8322.2012.00874.x/abstract.

Nancy, Jean-Luc. *After Fukushima: The Equivalency of Catastrophes.* Translated by Charlotte Mandell. New York: Fordham University Press, 2015.

Pringle, Thomas. "Photographed by the Earth: War and Media in Light of Nuclear Events." *Necsus: European Journal of Media Studies* 3, no. 2 (2014): 131-154.

Shiraiwa, Yoko. "Rescuing Tsunami-Damaged Photographs in Japan." *Journal of the Institute of Conservation* 36, no. 2 (2013): 195-203. https://doi.org/10.1080/19455224.2013.820205.

Sontag, Susan. *Regarding the Pain of Others.* New York: Picador, 2003.

Suwa, Nobuhiro, and Élise Domenach. "Filmer ce qu'on ne peut pas filmer." In *Dans l'œil du désastre: créer avec Fukushima,* edited by Michaël Ferrier. Vincennes: Thierry Marchaisse, 2021.

Takeda, Shimpei. "Trace – Cameraless Records of Radioactive Contamination." In *Making the Geologic Now,* edited by Elizabeth Ellsworth and Jamie Kruse, 206-213. New York: punctum books, 2013.

Tokyo Document Recovery Assistance Force. *Tokyo Document Recovery Assistance Force (blog).* Accessed September 12, 2025. https://toubunq.blogspot.com/.

Zylinska, Joanna. *Nonhuman Photography.* Cambridge, MA: MIT Press, 2011.

VERNON PRESS

Other distinguished titles from "Series in Critical Media Studies":

Hate speech and abusive behaviour on social media

A cross-cultural perspective

Luiz Valério P. Trindade

This book examines online hate and abuse on social media from a cross-cultural perspective, analyzing 108 studies in Italian, Portuguese, and Spanish. It reveals how culture and language shape harassment, with key insights for scholars and policymakers.

$53 | €49 | £42
Subjects: Media Studies, Communication, Sociology.
ISBN: 979-8-8819-0024-3 | Hardback | 132 pp | 9/2024
Also available in Paperback and E-book.
vernonpress.com/book/2062

Monsters and Monstrosity in Media

Reflections on Vulnerability

Yeojin Kim, Shane Carreon (Eds.)

This anthology explores how on-screen monsters reflect shifting notions of identity amid neoliberalism and the pandemic. Examining bodies in liminal spaces—human, machine, animal—it analyzes monstrosity as a site of subversion, vulnerability, and contested belonging.

$103 | €97 | £83
Subjects: Cultural Studies, Sociology, Media Studies.
ISBN: 978-1-64889-846-4 | Hardback | 218 pp | 3/2024
Also available in E-book and Paperback.
vernonpress.com/book/1926

Fix-It Fics

Challenging the Status Quo through Fan Fiction

Kaitlin Tonti (Ed.)

Fix-It Fics explores how fanfiction writers rewrite flawed narratives to challenge bias, heal trauma, and advocate for change. Focusing on the fix-it fic subgenre, this collection highlights fandom as a powerful space for activism, identity, and community building. It's an essential resource for scholars of pop culture, gender, and queer studies.

$103 | €97 | £83
Subjects: Sociology, Cultural Studies, Gender studies.
ISBN: 978-1-64889-814-3 | Hardback | 218 pp | 3/2024
Also available in E-book.
vernonpress.com/book/1894

Vernon Press is accepting proposals for monographs or multi-author volumes in this series.

For more information, visit https://vernonpress.com/publish-with-us or contact us directly at submissions@vernonpress.com

www.ingramcontent.com/pod-product-compliance
Lightning Source LLC
LaVergne TN
LVHW020414220926
848484LV00018B/2575/J

* 9 7 9 8 2 6 1 6 0 0 2 7 5 *